AF333791

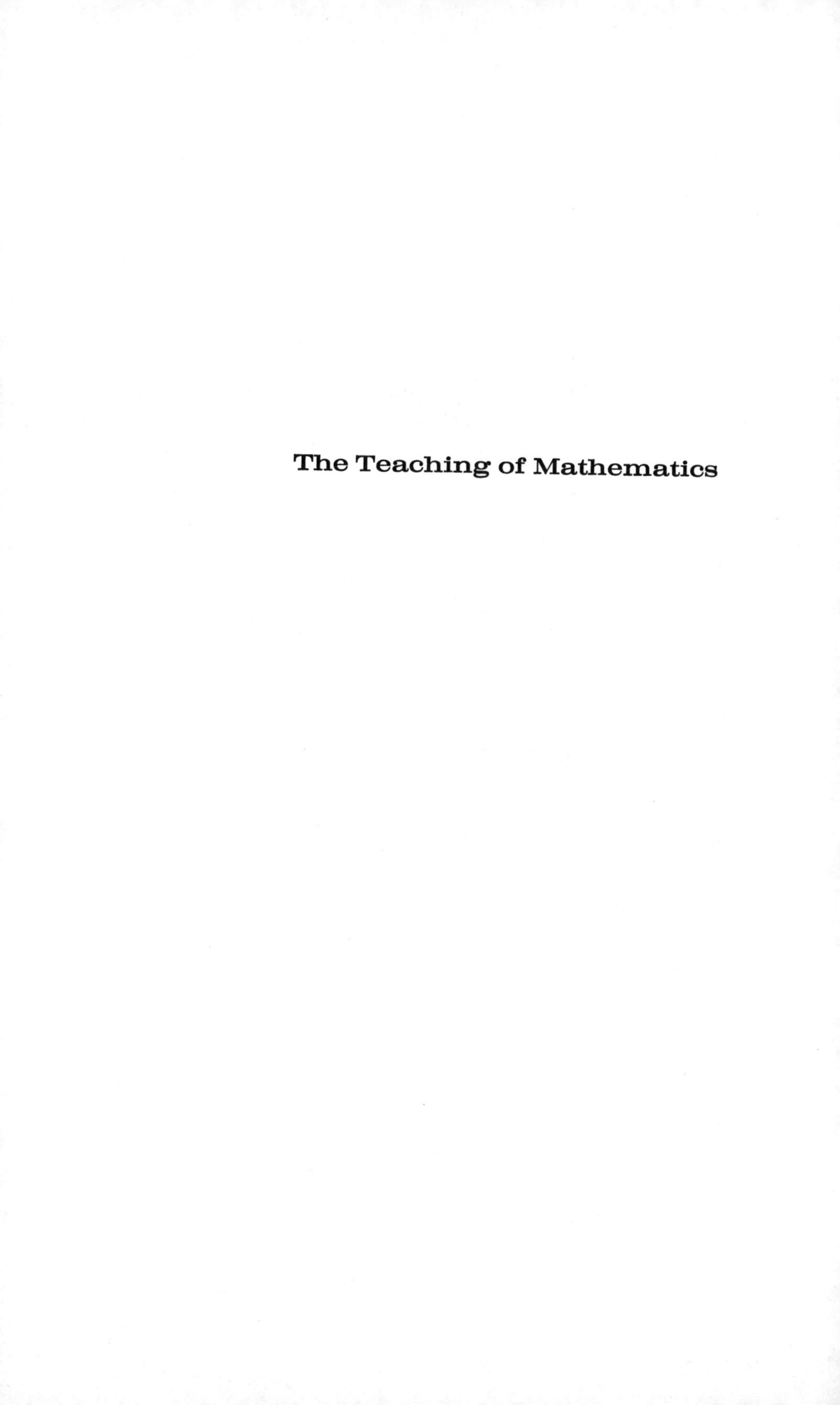

The Teaching of Mathematics

The Teaching of Mathematics

From Intermediate Algebra

through First Year Calculus

by Roy Dubisch

Professor of Mathematics, University of Washington

with the assistance of VERNON E. HOWES

Assistant Professor of Mathematics, Fresno State College

ROBERT E. KRIEGER PUBLISHING COMPANY
HUNTINGTON, NEW YORK
1975

Original Edition 1963
Reprint 1975

Printed and Published by
ROBERT E. KRIEGER PUBLISHING CO., INC.
645 NEW YORK AVENUE
HUNTINGTON, NEW YORK 11743

Library of Congress Cataloging in Publication Data

Dubisch, Roy, 1917-
    The teaching of mathematics.

    Reprint of the ed. published by Wiley, New York.
    Bibliography: p.
    1. Mathematics--Study and teaching. I. Title.
[QA11.D77 1975]   507'.11   74-23520
ISBN 0-88275-198-0

Preface

This handbook has been written with two main purposes in mind. The first is to provide the new teacher of mathematics with some general guidelines on the teaching of mathematics and some specific suggestions in regard to classroom procedures. The second is to provide both the inexperienced and the experienced teacher with an annotated bibliography of articles on the teaching of mathematics from the intermediate algebra level through first-year calculus.

A first draft of this book was written while I was a Faculty Fellow of the Fund for the Advancement of Education in 1951–1952, and I have frequently consulted this draft in my own teaching. In preparing a revision for publication, however, I, with the assistance of Professor Howes, not only have brought the bibliography up to date but have made a large number of changes and deletions. Most of these changes and deletions have been occasioned by the rapid changes in the teaching of mathematics during the last ten years. To cite just two examples: (1) until fairly recently, it was extremely rare to find, even in widely used calculus textbooks, any distinction made between a function and a functional value. The algebra textbooks all talked about "y is a function of x. . ." and thus never defined "function" at all. Now it is becoming fairly standard practice at the second-year algebra level or even earlier to discuss the concept of a function as either a mapping or as a set of ordered pairs. (2) Algebra, except at the junior or senior level in college, was never presented as an axiomatic system, and many college freshmen mathematics students had seen, at most, one or two brief mentions of the commutative, associative, and distributive axioms. Now we find that these axioms are being introduced even before high school, and that, by the twelfth grade, a fairly sophisticated use is often made of the axiomatic approach.

In particular, the publications of the School Mathematics Study Group (S.M.S.G.) are a rich source of the newer ways of presenting mathematics up through the twelfth grade (elementary function theory). Every instructor of mathematics at this level should certainly be familiar with their contents. Much of what I wrote in 1952 concerning the teaching of these subjects has been incorporated in these texts and the accompanying teacher's manuals.* Hence, for the subjects covered by S.M.S.G. texts, I have confined myself largely to remarks on common student difficulties and techniques for overcoming them, and to a listing of readings for teacher or student which either show different ways of considering the topic under discussion or provide additional information on the topic.

I definitely do not believe that there is just one style or method of good teaching. Although my opinions are presented along with those of others, each reader is urged to use or to ignore the given suggestions in the light of his own feelings about mathematics and education, as well as a consideration of the type of students he is teaching. What is important is that he realize that many people have given thought to the problems of teaching. By studying the various suggestions made, he will be able to adapt or modify them to fit into his own teaching procedures.

After three preliminary chapters dealing with general background material, the book goes directly into a detailed study of the courses which constitute the great bulk of our advanced high school and elementary college teaching. These chapters begin

---

*This mention of S.M.S.G. texts as a "model" is not to be construed as a blanket endorsement of the entire content of the series or to imply that other good or better text books are not available. But the S.M.S.G. text books are a good sample of the new approach and form a fairly coherent and easily available sequence for reference.

with general remarks about the subject under discussion. Then, following roughly the common order of presentation of the subject, the various topics constituting the subject are discussed.

No attempt has been made to provide a complete treatment of the teaching of this part of mathematics. All I am trying to do is to provide a handbook which will make the lot of the beginning teacher (and his students!) a little easier and which will provide the more experienced (and perhaps slightly jaded) teacher with food for thought and experimentation. Journal references are generally confined to the more easily available publications, and I have tried to be fairly selective. Certainly a large amount of trivial and unimportant material exists in the journals. For example, one journal devoted about ten pages (in four articles) to discussions of rules for determining the powers of i! A few of these references are included primarily for their historical interest. It is certainly interesting to observe that several eminent mathematicians of the previous generation advocated ideas that are just now being put into effect. The book references are by no means intended to be complete, and the few that are given are simply to illustrate procedures suggested in the discussions. Each reference is identified by a bracket [ ] referring to the bibliography at the end of the book.

With the exception of portions of the first and second chapters, this book is addressed to both the high school and the college teacher of mathematics at the level discussed. (The corresponding material of these portions of Chapters 1 and 2 at the high school level is certainly a part of the standard curriculum leading to a high school teaching certificate.)

ROY DUBISCH

Seattle, Washington
February 1963

Contents

7. The Teaching of Differential Calculus 81

8. The Teaching of Integral Calculus 96

1

Teaching

in General

Some of the problems of teaching are independent of the subject matter, and we discuss these first. That many books and articles have been published on teaching is attested by the size of the *Education Index* [E3]. The high school teacher will, of course, have had courses dealing with the general problems of teaching at the secondary level in connection with the obtaining of his credential. As for college teaching, the best books available for an overall picture are, in my opinion, [C14] and [M10]. A careful study of these books, together with a reading of the delightful and provocative book [B6], will give one an excellent picture of higher education. In fact, for the *beginning* teacher, the advisability of much additional reading on the general problems of education is highly questionable. There is so much controversy over basic issues that the novice can easily become quite confused if he seeks to settle all of the issues, and he may end up by losing his native talents as a teacher through attention to a mass of detail. Later on, of course, further study and investigation are desirable.

1. Objectives of Teaching

Before criteria of good teaching can be set up it is certainly necessary to describe the objectives of education. The objectives of teaching from the viewpoint of mathematics, alone, are described in Chapter 2. They represent, I think, generally accepted opinions. However, mathematics is only a part of the curriculum of any student, and the purpose of the discussion here is mainly to indicate criteria of good teaching in relation to the student's general education.

The interested reader can find statements of all kinds on the objectives of education, ranging from those which regard it primarily as a preparation for earning a living to those which regard it as being primarily for the inculcation of an appreciation for our heritage from the past. Thus, at one extreme, we have some engineering schools in which the curriculum is almost entirely technical, and, at the other extreme, we have colleges, such as St. John's, in which much of the curriculum consists of the study of "Great Books." Such broad considerations of educational policy, however, are not the concern of the beginner. He will enter a school in which a pattern already exists, and in the beginning his main concern will be with the teaching of the already existing mathematics courses. At some time in his career, however, he should certainly devote some time to the study of the general aims of education and the place of mathematics in it.

More important for our immediate purpose is the fact that the objectives for individual courses are often debated. Thus we find that some educators regard the primary objective of each course as developing an ability to think, while others feel that the imparting of information is the primary task. Still others advocate such nebulous aims as the inculcation of moral or spiritual values or an appreciation for democracy. Personally, I do not feel that the third class of objectives is attainable, while I would seek to combine the other two. Thus I believe that teachers should not try *directly* to teach students how to think, but rather should present information in such a way that the thinking involved in the subject stands out clearly, and the student can see the value of a thoughtful approach to the subject.

And, armed with facts plus an ability to analyze and handle facts, the student should draw his own conclusions regarding matters of morals and politics.

The question of the choice of facts is, of course, an important one and a legitimate subject for discussion, and much discussion is going on as in the School Mathematics Study Group (SMSG) for the elementary and secondary schools and in the Committee on the Undergraduate Program in Mathematics (CUPM) of the Mathematical Association of America. In this chapter, however, we are concerned only with the general problems of effective teaching.

2. Characteristics of the Good Teacher

Having defined a good teacher as one who has the ability to convey information in a logical and thought-provoking fashion, we now consider how this ability may be developed. While I hold strongly to the view that there are many types of effective teachers, and while I realize that no one type is best for all students, I think we should not let the great variety existing among good teachers blind us to the possibility of making some generalizations. For example, all of us can probably recall an almost totally unintelligible teacher who, by his very obscurity, forced us to work out everything ourselves. If we call him a great teacher, we are forgetting the large number of students who were unable to dig things out for themselves and hence never learned the subject. Or, on the other hand, if we call a great teacher a person who develops mechanical skills in his students to such a high degree that they can all pass an examination, we are forgetting that such a teacher has rarely inspired anyone to dig into the reasoning behind the mechanics.

There is widespread agreement among both teachers and students that a good teacher must know his subject and have an enthusiasm for it. In addition, however, he must have or develop the ability to

1. Present material in a thought-provoking way.
2. Explain clearly the reasoning needed to develop the subject and the technical skills necessary to apply this reasoning.

I say nothing here about the ways these ends may be accomplished, as a great deal of what follows is devoted to ways of attaining these objectives. In particular, I do not think it is important to argue for or against lectures, discussion, board work, etc. The use of these and other techniques will depend on each individual instructor, the type of course taught, and the students in the course. To make the objectives clear, however, I contrast the two of them by an elementary example:

It is possible to begin discussion of the derivative with a formal definition of the Δ-process and to explain very clearly the steps involved, so that every student can do all the problems of the homework. This presentation, then, satisfies criterion 2 but certainly not criterion 1. On the other hand, we can present an entertaining and highly stimulating discussion of the reasons for defining a derivative as we do (by, say, considering the velocity of a falling body) and yet, by passing off the mechanics with a casual remark and a few rapid maneuvers at the board, give the student no clear idea of how he can calculate derivatives by the Δ-process.

To retain a balance between these two important objectives is no easy task, and I suggest that, after each class session, the instructor evaluate himself on these two points. If, on the one hand, students showed eager attention throughout the class and asked questions but cannot do the homework, the teacher is probably neglecting the second objective. On the other hand, if they do the homework perfectly but are listless and bored in class and ask no questions, he is probably neglecting the first objective.

Much, much more has been written about good teaching, but all the advice I have seen can, I think, be boiled down to about this:

The good teacher is a human and mature person who knows his subject thoroughly, has a keen interest in it, and tries to get it across to his students in a thought-provoking fashion.

Such a general standard should serve the beginning instructor for at least the first two or three years, until he has had a chance to gather some experience. Then, I think, and not before, is the time for him to consider the whole question of teaching in a more systematic and detailed way. See, for example, [C14], pp. 557–594. Even then, however, he should remember that there is a large variety of methods of good teaching and that opinions vary on the criteria for good teaching.

3. Attitude toward Students

Some persons feel that a friendly personal attitude toward students is necessary for successful teaching, but this does not seem to be so—at least not for college teaching. Thus in polls of student opinion, for example, we do not find this quality as being considered important.* In the long run, however, teaching will become a bore if the instructor looks upon his students without personal interest. So whenever teaching begins to pall, try taking a more personal interest in both the poor students and the good students. There is considerable fun to be had in trying to get a difficult point through the thick head of a poor but hard working student, while the thrill of developing a good student is something that can really make teaching worthwhile. Recommended reading is [R22], which deals with student rating of instructors.

4. Examinations and Grades

Examinations and grades are a necessary evil and are certainly important to the student. As a beginning instructor you may be required to give only examinations constructed by others. If, however, you must construct your own examinations, I strongly urge that you ask the more experienced members of your department for copies of examinations they have given in the past, and model your first examinations after them. Like-

---

*That is, a teacher who is vitally interested in his subject but disinterested in his students can still do an effective job of teaching. I do not mean to imply, however, that a teacher who actually dislikes students can still do a good job. On the contrary, he is likely to be sarcastic and overbearing, and both of these attitudes are heartily detested by the students.

wise, in marking, it is best to find out the general custom of the department or school in regard to this matter, and follow it fairly closely at first.

Additional references for this chapter are [B32], [C4], [C23], [D11], [E1], [G3], [H22], [K9], [K10], [K11], [M17], [M23], [R5], [S4], [S32], [W21].

2

The Aims

of Teaching

Mathematics

In a discussion of the aims of teaching mathematics it helps to remember that there are, roughly speaking, four main classes of students of mathematics. These are:

1. Those who plan to take an advanced degree in mathematics.
2. Those who plan to teach elementary or high school mathematics.
3. The prospective majors in subjects that use mathematics extensively, such as engineering, physics, and chemistry. (We may also include in this group those mathematics majors in college who plan to enter an applied field such as computers after their baccalaureate degree.)
4. Those students who are taking mathematics as an elective either because of inherent interest in the subject (without planning to major in it or to apply it) or because of a college requirement for entrance or graduation.

In the larger universities there is often considerable segregation of these groups (especially the engineering students), but in the smaller colleges and in the high schools they are usually found together in many classes—especially the more elementary ones. Very frequently the special needs of one or more of these groups are sadly neglected. On the one hand, teachers sometimes treat a class of engineering students as if they were interested in applications alone and neglect those portions of mathematics which are of greatest interest to prospective mathematicians. Such an approach fails to realize that many prospective engineers have been attracted to the field because of its mathematical nature. Their interest is primarily in the field of mathematics, and with a realization of what real mathematics is like and how little of it is actually used in most engineering, they often change over into very fine mathematics majors. On the other hand, mathematics teachers frequently overlook the many fine opportunities that the applications of mathematics give for providing motivation and illustration of good mathematics.

1. Common Aims

At one time the cardinal virtue of mathematical training was alledged to be its role in the development of "mental discipline." According to this theory, the actual subject studied was not nearly as important as the mental discipline involved in its study. Under this banner the traditional curriculum of mathematics, Latin, rhetoric, etc., which was originally developed in the Middle Ages, continued to hold sway up through the nineteenth century. But under the onslaught of an army of twentieth-century psychologists the theory received severe punishment and, for a while, seemed to have died.

Present-day opinion seems now to recognize the existence of the possibility of some transfer of training. Thus in [S3] the author, after reviewing 810 references on the subject of transfer of training, says

It is my belief that there is enough in transfer of training to justify educators sticking to a rather conservative type of curriculum in which the hard subjects—Latin, mathematics, science, history, modern languages—still find a place, at least for the more intellectually-minded of our pupils. Nowhere in transfer experience is there evidence supporting the theory of Progressive Education that the selection of subject matter according to the whims and fancies of an immature human being results in a beneficial transfer, and therefore, in education. We ought to distinguish between work and plan in education, for it is through work that transfer takes place.

The logical habits of thought used in mathematics *can* be transferred to a limited extent to other subjects, but only if mathematics is taught in such a way as to emphasize the possibility of such transfer. That is, the old theory, which maintained that the study of mathematics, as an entity in itself, would promote habits of logical thought in any other situation, is no longer held. But, if the fact that the logic used in mathematics is applicable everywhere is stressed in the teaching of mathematics, and if examples are given of how this logic applies, there may be considerable transfer. See, for example, [P10], [H15], and [Z1].

There is an enormous amount of literature on the topic of transfer of training. See, for example, [B22]. I do not give further details here, since I feel that the importance of the issue has been greatly overrated and that the defense of mathematics as an item in the curriculum can be made without recourse to the transfer of training possibilities.

Regardless of the possibility of the transfer of training, most teachers agree that *mathematics should be taught with emphasis on the thinking process.* (See, for example, [N9].) A considerable part of the material of this book will be devoted to methods of increasing this emphasis. It is all too easy for teachers, experienced and inexperienced alike, to lose sight of this primary aim in the struggle to get students to pass examinations in routine material. One extreme case has been reported in [R17] (on the high school level, it is true) of a New York teacher who found that students in trigonometry had most trouble in the state Regent's examinations with the logarithmic solution of oblique triangles. She therefore *began* her course with a thorough rote treatment of this topic!

It is also true that many teachers, while recognizing that

emphasis on the thinking process is appropriate in the training of prospective mathematicians, are inclined to feel that for engineers, for example, a routine mechanical-like skill is sufficient. That this is not so, is amply testified for by the engineers themselves (e.g., see [M27]). In this article, by the way, the author asserts that how we teach is of overwhelmingly greater importance than what we teach, in the sense that it is the basic methods of mathematical analysis that are applied, rather than the detailed results. In a similar vein a report in [H27] on a conference on training in applied mathematics concludes that: "The student should receive a broad, thorough training in *fundamental* mathematics . . . " [my italics]. See also [H31], [K5], [R16], and [S36].

Having stated the primary aim of all mathematics teaching—the emphasis on the thinking process—we are now ready to discuss special needs. In so doing we need to keep in mind, of course, the fact that these needs are not completely separate and that students may change from one classification to another.

2. The Needs of the Prospective Mathematician

Those readers who are in the process of completing their graduate training, or who have recently done so, will most clearly recognize the inadequacy of many present-day undergraduate programs as a preparation for graduate work. It is true that the taking of almost any mathematics course (if taught with an emphasis on the primary aim) will contribute to the development of that intangible but vitally important attribute known as mathematical maturity. Nevertheless, except for that intangible gain, much of the traditional program has been a waste of time from the standpoint of modern graduate mathematics. See, for example, the CUPM recommendations [C17] for a "modernization" program for prospective teachers of mathematics and the forthcoming report of its Panel on Pre-graduate Training.

This common dichotomy between undergraduate and graduate mathematics can produce some very bad results. It has actually happened that students who have done very well in "traditional" undergraduate mathematics do very poorly in graduate work, and, vice versa, students who have done poorly in undergraduate work (through lack of interest, of course, rather than through lack of ability) have done superbly in graduate work. On the one hand, then, we have students who have every reason to expect success in mathematics, meeting with a cruel disappointment, and, on the other hand, we have the loss of many prospective good mathematicians because they fail to see much that is really interesting in mathematics.

At one time, of course, this dichotomy did not exist. The traditional course in theory of equations, for example, was excellent background for the "modern" algebra of the time, with its work on invariants, covariants, and the like. The modern algebra of today does not even require a course in theory of equations for comprehension, and the student who expects "more of the same" when he advances from one to the other is in for a surprise.

In general, the current reforms in the curriculum emphasize that the main objective in the teaching of mathematics for the prospective creative mathematician is to emphasize the *abstract nature of mathematics*.

Note that this is not quite the same as our primary objective— the emphasis on the thinking process—although it is related. For example, a student may follow with keen interest a *manipulational* proof and thereby display the desired commendable interest in thinking, but be uninterested or unable to follow an abstract argument. The classical example of the difference here lies in the history of invariants. At one time results in this field were obtained by ingenious calculations, and P. Gordan was crowned "King of Invariants" for his ability to excel in this sort of thing. But his brute-force proof of the existence of a finite fundamental system of invariants and covariants for systems of binary quantics was replaced by Hilbert's noncomputational basis theorem, and it was Hilbert's reliance on general reasoning and existence proofs, rather than calculation, that was said to have caused Gordan to complain, "This is not mathematics; it is theology." (See [B11].)

There is another aspect of mathematics so commonly neglected that it deserves mention here. This is the philosophy of mathematics, together with the attendant topic of logic. I have a strong feeling that there are quite a few students who would be interested in this phase of mathematics if they knew about its existence. As it is, such students frequently end up in the philosophy department and study philosophy without what I regard as the necessary mathematical background. From time to time references to the philosophy and foundations of mathematics should be made and interested students referred to such books as [R13], [R30], [T1] and the articles [B27], [W7], and [W13].

3. The Needs of the Prospective High School Teacher

I often feel that the students of this group are the "orphans" among our mathematics majors. Generally speaking, this group consists of students who range in ability from slightly better than average to good in the traditional (relatively nonabstract) type of mathematics. Since they do not, in general, make good prospective Ph.D. candidates they are often not considered as being "real" mathematics majors. (Majors in other fields often get this neglect from their mathematics teachers too, but they are treated with proper respect in their own respective departments!) As a result, all too frequently these students go out to the high schools to become narrow-minded purveyors of mathematics as manipulation (at which they themselves excel). Or, worse yet, they flee from the chilly halls of mathematics to the warm embrace of Education (note the capital E) and become advocates of Method versus Content.

To save these students so that they may become badly needed centers of strength in the high schools should be an important objective on the part of every college mathematics teacher. To do this he should first learn to treat these people as prospective allies in a common cause and not as second-raters. It is true that they often do not have the ability in abstract reasoning that a college teacher presumably has, but they may have other equally rare qualities that are badly needed in the difficult task of high school teaching. See, for example, [K14] and [K15]. Second, the college teacher should hold that:

The prime objective in the teaching of mathematics to prospective high school teachers is to get them to see the bearing that advanced mathematics can have on their high school teaching.

Too often, topics discussed in undergraduate work are considered mainly as steppingstones to further work, and no reference is made to their bearing on past work. The example of a very eminent mathematician, Felix Klein, should convince anyone that the backward point of view can definitely be made worthwhile. His classic work [K19] should be required reading for every mathematics teacher. Another book with the same objective and better adapted to our present U.S. high schools is [F2]. Further references on this topic are given in [B18], [E6], [H13], [J9], and [Q1]. A major task of CUPM is that of developing a curriculum most suitable for the prospective elementary or high school teacher of mathematics. Every college mathematics teacher who has any contact with prospective teachers of mathematics in the elementary or secondary schools should read the CUPM reports [C16, C17] on this problem.

4. The Needs of the Prospective Scientist

On the whole, I would say that, up to recent times, the needs of this group have been the best met among the four groups mentioned. This is not to say, however, that perfection has been attained. On the contrary, bitter criticism has been made of mathematics teaching by teachers of subjects in which mathematics is used as a tool. This criticism, however, consists less of complaints about students not knowing how to perform mathematical operations than it does of complaints that students cannot *apply* their mathematics to the subject being studied. That is, transfer of training is not taking place sufficiently well here.

Our objectives for this class of students should be

1. To teach general principles so that they may be applied in different situations, rather than to teach mechanical schemes which apply only to (often artificial) textbook situations.

2. To state clearly and honestly where and to what extent various mathematical processes are used in the sciences.

Although our purpose here is simply to state objectives rather than to discuss how they may be attained, a sample of methods used to achieve the first objective may help to make it better understood. For example, in integral calculus it is much better, from the standpoint of developing the student's ability to use calculus easily and naturally in his science courses, to emphasize the heuristic idea of summing up "little" elements to get at a problem than it is to stress formulas for the separate cases of area, volume, etc. This does not mean a neglect of such things as Duhamel's theorem (or equivalent) but an emphasis on an intuitive "practical" approach so that the science student will develop a "friendly" feeling toward integral calculus. He should come to look upon it as a natural and easily used tool for solving difficult problems rather than as an esoteric device to be but vaguely remembered on the job.

The second point is often neglected because of lack of knowledge on the part of the instructor. One of the purposes of this book is to help remedy this deficiency. Thus when we discuss the various topics of mathematics, comments will be made about their use as "often used," "seldom used," etc. To avoid later misunderstanding, I want to emphasize now that these statements are only generally true and apply only to the "average" student. Thus it is quite conceivable that a student who goes on into mathematical physics will find extensive use of a certain topic for which little use is found by the average engineer or scientist. Collections of problems using mathematics in applications are given in [M5] and [S31]. For further references on the subject of applied mathematics see [C15], [C20], [D4], [D5], [D20], [E5], [G2], [G8], [H8], [H12], [H19], [H29], [H32], [J6], [K3], [K5], [L14], [M18], [N6], [R14], [S30], [W8], [Z2], and, in particular, the report of CUPM [C18] on the undergraduate mathematics program for engineers and physicists, where it is pointed out that the engineers and physicists of today have an urgent need for many types of mathematics—for example, matrix algebra, linear and nonlinear programming, Monte Carlo methods, etc.—that are not a part of the traditional program for these students.

There is one other aspect of the training of scientists that I want to mention here since it is frequently neglected by both mathematicians and scientists. This relates to the connection between mathematics and reality. I believe it is very important that prospective users of mathematics understand this point very clearly. Too often, I feel, scientists have a blind faith in the power of mathematics and look upon it as, in itself, an empirical science. Since such a viewpoint can lead to considerable confusion in the advanced phases of any science, I strongly urge that the nature of the real relationship between mathematics and science be presented at some time in the prospective scientist's study. For excellent discussions of this point see [B15], [B26], [B28], and [H18].

5. The Needs of the Liberal Arts Student

The aims and objectives of the teaching of mathematics from the viewpoint of the liberal arts student, who is not majoring in either mathematics or the sciences, have been the subject of very considerable discussion and controversy. Broadly speaking, most writers on this subject want to favor the fundamental principles of mathematics as opposed to the mechanics and to explain the nature of mathematics and its contribution to our civilization. It is on the question of the level of presentation and the topics to be discussed that we have such a divergence of opinion. Since there seem to be about as many points of view on this matter as there are writers, there is little point in trying to summarize matters here. As samples of such writings the reader may consult [A5], [C13], [D8], [N7], [N8], [O7], [R26], [S8], [S18], and [T10].

Additional references for this chapter are [B10], [B16], [C1], [D16], [G6], [R7], [W10]. [R7] contains statistics on courses actually offered at 113 four-year colleges in 1955. A forthcoming CUPM report will provide more up-to-date information.

3

General Remarks

on the Teaching

of Mathematics

We present in this chapter a discussion of certain aspects of teaching which apply about equally well to all mathematics courses—from intermediate algebra through the elementary calculus—but which were not covered in the general discussion of Chapter 1.

1. Homework

It is generally agreed that homework should be assigned at every class meeting (except when a test is given at the next class meeting). The well-known rule of two hours work outside of class for every hour in class could be applied to gauge the amount of homework assigned. In practice, however, three hours for the average student seems to be much more common for mathematics classes. If homework requiring more than this amount of time is assigned, a large portion of the class will not be able to complete their assignment and will become discouraged. Since problems vary in difficulty, it is impossible here to translate the hourly requirements into a certain number of problems. The beginning teacher can consult his colleagues (being prepared to get a variety of answers!) and he should check with his students from time to time to see how much time they are spending. A rough estimate can usually be obtained by the instructor if he works the problems himself and multiplies the time it takes him by four (this conversion factor, of course, is only approximate and will vary from instructor to instructor, as well as varying for each instructor according to his experience).

Although it would be a waste of the instructor's time to work through every homework assignment in detail, I do suggest that the beginning instructor, at least, will find it worthwhile to glance over the homework assignment and make at least a few notes on some of the problems that look as if they might cause difficulty. Often a trick is involved in a problem, and it may readily escape his mind when in front of a class; and nothing can be more embarrassing and more detrimental to class morale than to have the instructor unable to do a problem he has expected the class to do.*

Some statements should be made to the class concerning homework. A very common failing of students is to attempt to work their homework problems without any preparation other than the class lecture. If you can convince your students that 15 to 30 minutes of careful study of their text and notes *beforehand* will often enable them to do an assignment in an hour, which might otherwise take them more than two hours, you will have gained a major victory. Second, it is wise to emphasize the fact that too much time can be spent on any one homework problem. Very often a student will get stuck on some minor point and will waste hours on it that could be better spent in doing other problems. Third, you should make very clear to the class your policy toward the form in which papers must be handed in, and toward

---

* An occasional bit of trouble along this line is nothing to worry about. A frank admission of temporary inability, however, is better than an attempt to bluff.

late papers. Furthermore, it is only fair to state just how much homework counts (if any) toward the final grade.

On these last three points there is considerable difference of opinion. Perhaps your department will have a policy which they expect you to follow. Check on this. If not, and if you are marking your own papers, I think it will pay to insist on homework being done in some standard form on standard-size paper, with no late papers allowed except in cases of excused absences. (If you have a reader, he would appreciate the same policies!) On the last point we find variations, ranging from those instructors who count the homework for as much as 25% of the grade to those who do not count it at all. Personally, I keep a record of the homework grades for the purposes of counseling, but count homework only when there is a doubt as to whether the grade should be "D" or "F." In such a case, I would give a "D" to the student who has consistently handed in his homework and an "F" to one who has not.

If credit is given for homework in the final grade, there is constant pressure for students to copy, to challenge homework grades, and to plead for submission of late papers. Furthermore, the better student will often waste his time working routine problems when he could spend his time more profitably doing more difficult problems or outside reading, and the poorer student will spend excessive time, to the detriment of his other courses, in trying to do all his homework. On the other hand, it might be argued that, without the incentive of credit toward a grade, students will not do their homework. Generally speaking, I do not find this so. The great majority of students realize the importance of doing their homework as a preparation for tests* and do as much of it as they would if a grade were assigned. Very occasionally a student will ruin his chances for a passing grade by not doing homework, but this usually occurs only once, and I feel that he has learned a very valuable lesson in this case.

It is generally agreed that, whether or not homework is used

---

* Particularly if the tests, as I feel they should, emphasize the type of problems encountered in the homework. An occasional test problem taken directly from the homework will often make students take a more active interest in their homework.

for credit, it should be collected, graded, and returned promptly to the student at the next class meeting. And, when it is returned, students should be urged to make the necessary corrections and to keep their papers for future reference. If an instructor is offered a reader for his papers, it is not likely that he will refuse him. Nevertheless, there is considerable merit in grading at least some of your own homework papers, especially when new at the job, since there is no better way to find out students' difficulties. If you do use a reader, be sure to instruct him to give you a report on what seem to be the general difficulties in the assignment. (With experience, of course, you will usually know them in advance.)

Too much time should not be spent in grading homework, by the teacher or the reader. An hour for an average assignment for a class of thirty should be sufficient. Pick out the paper of one of the better students, and after going over it fairly carefully, use it as a key. If a problem is found incorrect, mark an X at the point where the first mistake occurs. Do not try to give an exact grade, but use, say, 0 to 10 as a scoring basis. Make notes of the difficulties commonly encountered for class discussion.

One more problem in connection with homework needs to be discussed. On the day the homework is due there will usually be questions on some of the problems. If the instructor simply asks for questions on the problems he is all too likely to wind up doing all of the problems and not finding time for new work. A poll to find out what problems caused the most difficulty is likely to prove too time consuming. Many instructors, therefore, will not discuss the homework until after it is graded, and they know what problems caused the most difficulty. To save time in discussing these problems, they may write on the paper of a student who has done the problem in question correctly, "Please put problem — on the board." (Generally there are few problems that some student has not worked.)

This method has the disadvantage of letting the problems get "cold" before discussion—especially if the class meets only two or three times a week. An alternative method, which I have found useful, is to ask the first student who comes into the room to list the homework assignment on the board. Then each student coming in has the privilege of "voting" for, at most, three problems with which he has had difficulty. The three problems having the most "votes" are then discussed immediately at the beginning of class (with the provision that a minimum number of votes—say five for a class of thirty—is required for discussion). This system may be abused at first, but abuses are quickly detected, and it then works very well. Whichever method is used, some discussion of homework should certainly be given if the assignment has caused any general difficulty.

2. Choice and Use of Textbooks

The textbook is usually prescribed for beginning instructors, but at some time or another every instructor is faced with choosing one, either alone or as a member of a textbook committee. In selecting the book, of course, there is not a problem of scarcity but one of superabundance.

Your best source of information for elementary textbooks lies in the reviews published in the *American Mathematical Monthly, Mathematics Magazine, Mathematics Teacher, School Science and Mathematics,* and *The Mathematical Gazette* (published in Great Britain). These should be evaluated, of course, in the light of your own views and the needs and abilities of your students. Your own views need to be considered in order for you to use the book effectively, but do not forget that the needs of the students come first. You will, for example, undoubtedly prefer to teach from the most "high-powered" book you can find, and, if your students can take it, this is fine. But if they cannot (as is more likely!) both you and they will find it tough going.* They will blame you for the choice of the textbook and for your inability to make the text understandable.

Occasionally you will find yourself saddled with a textbook which contains erroneous material. Try to get it changed. If you cannot, do not hesitate to criticize the book, but beware of

* This is not to be construed as an argument for choosing a poor book because it is easy. But, for example, a completely postulational approach to the real number system is certainly not suitable for a beginning algebra course.

discarding it entirely in favor of lecture notes. Most students in high school and in the first two years of college are unable to take lecture notes on mathematics successfully, and unless the lecture or discussion follows the textbook fairly closely, they will be lost. On the other hand, of course, you should be careful not to give the impression that the book is the final authority and that the only way to do mathematics is to consult a textbook. Certainly the students should be convinced that an elementary textbook is merely a convenient reference work and that, generally, the logical demonstrations given there are not original with the author.

3. A Central Problem of Mathematics Teaching

One of the outstanding problems in the teaching of mathematics is the question of level of rigor to be used. It is certainly true that complete rigor is unattainable in elementary courses.* That is, for example, no one to my knowledge has suggested teaching beginning calculus on the real variables level.

The exact level of rigor will depend, of course, on the type of students in the class. The general approach that should be used, however, has been nicely expressed in [H17]. These authors state that the main ingredients of good mathematics are relevancy, rigor, and elegance, and they claim that it is possible to serve such a dish to the undergraduate. The main way to achieve this objective is to " . . . *pack the hypotheses* with carefully chosen assumptions that have the power to push through the proof in a rigorous and striking manner and yet are either obvious to the student or acceptable on the basis of past experience."

One of the examples they give of this procedure is as follows:

Prove $\dfrac{d(\log_a x)}{dx} = \dfrac{c}{x}.$ As our hypotheses we take

$$(1) \qquad \frac{d(\log_a x)}{dx} = g(x) \text{ exists for } x > 0$$

* Overlooking the fact that there is considerable controversy among mathematicians as to what constitutes complete rigor!

and

$$(2) \quad \log_a x + \log_a \alpha = \log_a \alpha x \text{ for } x > 0 \text{ and } \alpha > 0.$$

Differentiate the left-hand side of this equation regarding α as a constant, and obtain, by (1), $g(x)$. Differentiate the right-hand side with the help of the chain rule for the derivative of a function of a function and obtain $\alpha g(\alpha x)$. Now let $x = 1$ to obtain $g(1) = \alpha g(\alpha)$ or $g(\alpha) = g(1)/\alpha$.

Likewise the real number system can be presented on many levels of rigor, ranging from the Russell-Whitehead formulation, where everything is built up from concepts of logic, to a formulation which accepts most of the rules about numbers as axioms. As long as we are true to our own set of assumptions, we are not abasing mathematics. As E. H. Moore said, "Sufficient unto the day is the rigor thereof." In this connection see [B14] in which the author argues for the inclusion of some material concerning the doubts and uncertainties in the foundations of mathematics. See also [K13] and [W10].

Another point closely related to the one above has been stressed by many writers. They point out that, frequently, "proofs" are merely *verifications* of unmotivated, dogmatic statements. For a discussion and examples of this point see [M13]. The author gives, among others, the examples of the introduction of $e^{\int P\, dx}$ for an integrating factor, $y = e^{mx}$ as a solution of a linear differential equation, and the function commonly introduced in the proof of the law of the mean from Rolle's theorem. He urges that " . . . each proof and derivation be well motivated, natural, and therefore more acceptable to the student." An outstanding exponent of motivated and heuristic presentation is G. Polya. In talks, articles, and books he has constantly stressed the importance of a less formal approach to mathematics. See, for example, [P6], [P7], [P8], [P9], [P10], and [P11].

Along similar lines, in [B8], the author pleads for a consideration of the psychological approach to a subject to begin with, rather than the logical approach. For example, he suggests that we begin calculus with a problem in maxima and minima rather than with formalized limits and, in trigonometry, that we begin with acute angles and right triangles rather than the general angle. He urges that less attention be paid to "pathological" detail in beginning any course and says that the teacher should pick out something to motivate the student to start with and bring in details later.

4. Methods of Presentation

The most valuable study yet made on the teaching of college mathematics is, in my opinion, [S22]. The author's recommendations are based on the observation of 150 classes in elementary college mathematics. While he recognizes that no one method of teaching is best for all teachers, all students, and all topics, he presents the following valuable analysis of seven types of teaching, arranged in order of what he considers ascending value.*

(Page references refer to [S22].)

1. Type v—A ground-covering, textbook repeating recitation, in which students participate technically and mechanically. Placed lowest because:
   (*a*) It lacks motivated activity: teacher's questions, student's questions, teacher's answers, student's answers, rigid adherence to the textbook, and the teacher's development of the lesson, are all characteristic of a not too model, dead organism.
   (*b*) It lacks clear-cut exposition.
   (*c*) It lacks organization.
   (*d*) The teacher serves little purpose other than that of being a drill master ("master" in a questionable sense).
   (*e*) It fails to differentiate between the fundamentals and the details of the course.
2. Type vii—A blackboard recitation, at which the students recite and get a grade.
   (*a*) Exemplifies openly-arrived-at, individual, easily evaluated student activity.
   (*b*) At its best is a desirable laboratory exercise.

* Reprinted by permission of the Bureau of Publications, Teachers College, Columbia University.

(*c*) As ordinarily conducted is merely a traditionally accepted scheme of obtaining a student's grade.

(*d*) Is considered by many teachers a necessary evil which may continue in vogue until college teachers of mathematics learn to use, and agree to adopt, the modern "objective" testing.

3. Type i—The lecture which goes far afield, generally instructive and entertaining. The students, however interested, remain distinctly passive.

(*a*) Has the well-known faults of lecturing (the "pouring in" process) as a mode of teaching.

(*b*) Fails to reach the lower (in intelligence) half of the class.

(*c*) Precludes student activity.

(*d*) Is usually well motivated, well organized, rich in applications and contacts with other fields, and well prepared.

(*e*) Is valuable to the extent of the teacher's ability to "cast a glamour" over the subject matter.

4. Type vi—A carefully planned question-and-answer development.

(*a*) Is well organized and carefully prepared.

(*b*) Is very definite in planning its objectives.

(*c*) Is very successful in achieving its aims.

(*d*) Leaves the student with a very definite sense of personal accomplishment.

(*e*) Offers no opportunity for student initiative.

(*f*) Produces an undesirable amount of nervous tension.

(*g*) Weakens (certainly does not develop) the student's ability to cope with a whole problem.

(*h*) Is devoid of "higher" aims.

5. Type ii—The lecture used as a method of developing a theory, or a problem, or an exercise. . . . The students . . . seem to participate "actively" in the proceedings.

(*a*) Creates, and usually sustains, genuine interest.

(*b*) Demands, and usually holds, student's attention.

(*c*) Is uniformly characterized by clear, specific, and pointed "explanations."

(*d*) Depends for effect on the teacher's ability to dramatize the series of events leading up to the solution of the problem.

(*e*) Depends on the teacher's ability to anticipate student's questions and, more generally, on the teacher's intimate knowledge of the abilities and the interests of his students.

(*f*) Has the shortcomings (*a*) and (*c*) of type i.

6. Type iv—A combination of blackboard and oral recitation. The students show a lively interest; the instruction is casual and entertaining.

(*a*) Is a fine blending of the more desirable characteristics of types ii and vii.

(*b*) Is well adapted to create, in the students, genuine activity and alert responses.

(*c*) Encourages student initiative.

(*d*) Is very efficient, yet pleasantly informal.

(*e*) Obtains students' spontaneous reaction, by relegating grading to the background.

7. Type iii—The recitation at which the time is fairly evenly divided between instructor and students. The students' questions appear to be spontaneous. The instructor is resourceful and enthusiastic.

(*a*) Most nearly approximates the better qualities of essential determinants of teaching [which are presented on page 70].

(*b*) Contains the least number of violations of the principles of learning [which are presented on page 54].

(*c*) Places comparatively little emphasis on examinations.

(*d*) Is not dominated by the textbook.

(*e*) Utilizes the educative possibilities of the question to a fuller extent than any of the other types.

Examples are given of all of these various types of teaching.

5. Common Errors in Mathematics Teaching

The outline presented in Section 4 reveals a number of weaknesses in various methods of presentation. In the book from which the outline was taken a number of other common weaknesses are pointed out and illustrated by actual classroom incidents. Some of them are

1. Failure to realize that too many facts should not be presented at one time. For example, one teacher in a calculus class attempted to dispose of partial fractions by assuming the algebraic theorem on the decomposition of partial fractions and treating rapidly a single complicated example:

$$\frac{x^5 + 2}{(x - 2)(2x + 3)(x + 4)^2(5x^2 - x + 7)}.$$

2. Failure to realize that training in reasoning is rather specific, so that reasoning cannot be done in any field without experience in that field. For example, in a college algebra class, a teacher expected his students to reason out problems in permutations and combinations without letting them become familiar with the concept of a permutation through simple examples.
3. Failure to realize that skills should not be acquired a long time in advance for uncertain future needs. For example, an instructor in an algebra class spent three class periods in determinant evaluation without a word about their use, their application, or their *raison d'être*.
4. Failure to provide incentive and motivation. (There is, unfortunately, no need to provide examples of this!)

Another common error pointed out in [O4] is a failure to realize that most students are very weak on vocabulary. Thus, a survey by this author of 60 freshmen at the beginning of the college year showed that **33** did not know what a polynomial was; **43** defined term incorrectly; **11** missed quotient; **20** went astray on coefficient, etc.

6. Miscellaneous Remarks

1. On the question of the methodology of problem solving see [P6], [H2], and [B29]. The first of these references is directed mainly at the student, although it has much material of value to the teacher. A summary of the book is given at the end as follows:*

* Reprinted from *How to Solve It* by G. Polya by permission of Princeton University Press. Copyright, 1945 by Princeton University Press.

Understanding the problem

First.
You have to *understand* the problem.

- *What is the unknown? What are the data? What is the condition?*
- Is it possible to satisfy the condition? Is the condition sufficient to determine the unknown? Or is it insufficient? Or redundant? Or contradictory?
- Draw a figure. Introduce suitable notation.
- Separate the various parts of the condition. Can you write them down?

Devising a plan

Second.
Find the connection between the data and the unknown.

You may be obliged to consider auxiliary problems if an immediate connection cannot be found.

You should obtain eventually a *plan* of the solution.

- Have you seen it before? Or have you seen the same problem in a slightly different form?
- *Do you know a related problem?* Do you know a theorem that could be useful?
- *Look at the unknown!* And try to think of a familar problem having the same or similar unknown.
- *Here is a problem related to yours and solved before. Could you use it?* Could you use its result? Could you use its method? Should you introduce some auxiliary element in order to make its use possible?
- Could you restate the problem? Could you restate it still differently? Go back to definitions.
- If you cannot solve the proposed problem try to solve first some related problem. Could you imagine a more accessible related problem? A more general problem? A more special problem? An analogous problem? Could you solve a part of the problem? Keep only a part of the condition, drop the other part; how far is the unknown then determined, how can it vary? Could you derive something useful from the data? Could you think of other data appropriate to determine the unknown? Could you change the unknown or the data, or both if necessary, so that the new unknown and the new data are nearer to each other?
- Did you use all the data? Did you use the whole condition? Have you taken into account all essential notions involved in the problem?

Carrying out the plan

Third.
Carry out your plan.
{ • Carrying out your plan of solution *check each step.* Can you see clearly that the step is correct? Can you prove that it is correct?

Looking back

Fourth.
Examine the solution obtained.
{
• Can you *check the result?* Can you check the argument?
• Can you derive the result differently? Can you see it at a glance?
• Can you use the result, or the method, for some other problem?

In the second reference we have a treatment of somewhat the same topic from the point of view of the research mathematician. The third reference discusses the strategy of mathematics as often consisting of changing the form of a problem as, for example, (1) partial to comprehensive relations (as special word problems in algebra yield to a general approach); (2) simplification (as from infinite to finite when we replace limits by ϵ's and δ's); (3) generalization; (4) from exact to approximate; (5) relativization; (6) logical changes in form (as from a proposition to its contrapositive); and (7) integral to differential and conversely.

2. On the problem of teaching the engineering student see [A11] and [C6]. Much of the material contained in these articles is given in appropriate places in the following chapters.

3. Frequent exhortations are addressed to the teachers of mathematics to enrich their courses with historical material. Too frequently, however, the statement of minute, dry-as-dust historical facts is considered enriching. Thus we find, for example: "The calculus was invented independently by Isaac Newton (1643–1727) and Gottfried von Leibniz (1646–1716)," accompanied by two dreary-looking pictures of the gentlemen in question. Such a presentation contributes very little to an appreciation of either history or mathematics. It is far better, in my opinion, to include comparatively few items of history, but to discuss these items thoroughly. The human side of mathematics should be emphasized and the creators of mathematics presented as real people. Ample material for this type of presentation may be found in such books as [B11].

4. Students having difficulties with mathematics should be urged to read [A10], where suggestions on how to study mathematics are given under five headings: (1) general instructions, (2) how to study the text, (3) how to solve problems, (4) how to retain what is learned, and (5) how the teacher will help. [D1] is also an excellent little pamphlet on how to study mathematics. Finally, in connection with student difficulties in general, there is an article [J3] in which the author finds, after a long investigation, that " . . . self-confidence and work are the necessary and sufficient requirements for satisfactory scholarship in college mathematics"!

Additional references for this chapter are [B5], [B39], [C2], [C5], [D2], [G7], [H10], [H14], [J4], [J10], [K20], [L14], [M21], [N2], [O3], [S25], [S26], [T7], and [V2].

For an extensive bibliography dealing mainly with high-school mathematics and the first year of college mathematics see [S7].

4

The Teaching
of Algebra

Since elementary algebra, intermediate algebra, and college algebra (and even theory of equations) overlap to a considerable extent, we present algebra as a unit, in order to avoid duplication. The reader should realize that some of the remarks made apply more to intermediate algebra than to college algebra and vice versa.*

1. General Remarks

Before we proceed to a topical discussion, there are a few general points concerning the teaching of algebra which can be made. First of all, although the problems of the teaching of algebra in college are by no means the same as those of teaching it in high

---

* The problems of teaching beginning algebra are considerably different from those of advanced algebra, and we do not consider these problems here. See, for example, the SMSG first-year algebra text and teacher's commentary [S14].

school, there are definitely points in common. Hence a reading
of one or two of the books on the teaching of algebra in high
school is recommended for college teachers of second-year algebra. See, for example, [D6], [K18], or [R12] for traditional
approaches and [M12] or [N5] for more modern approaches.

Second, if there is any one point on which teachers of algebra
agree, it is that the numerical side of algebra should be stressed.
That is, the rules and methods of algebra should always be first
explained from the numerical viewpoint. It is foolish, for example, to expect students to add algebraic fractions when they
do not thoroughly understand the mechanism behind the addition
of $\frac{1}{2}$ and $\frac{1}{3}$.

Third, while both intermediate algebra and college algebra
must necessarily contain considerable review material, it is sometimes deadly to the better student to take all the review material
at first. It is perfectly feasible to mix review material with new
work, coming back to review whenever needed.

2. The Axioms of Algebra

Central to the problems of the teaching of algebra is the question
of the role of the axioms. Unless axioms are carefully stated
(along with the necessary definitions) it is difficult to give proofs
of such identities as $(-a)(-b) = ab$, $-(a+b) = -a-b$, etc.
The successful use of the SMSG texts, [S16] and [S17], certainly
proves that a fairly rigorous axiomatic treatment of the real
number system is possible at the intermediate and college algebra level.

Here, however, is certainly a case where the textbook used is
bound to have a very determining effects on the level of presentation. With the SMSG publications as a guide, we leave it to the
individual instructor to decide just how much of this approach
he can adapt to his particular class. Some things he can certainly do. For example, the fact that $3a + 2a = 5a$ does not
need to be "explained" by saying that "To add two similar mono-
mials, we add the numerical coefficients and prefix this sum to
the literal (letter) coefficients." It is certainly far preferable to
say, "Use the distributive law."

There are many excellent treatments of the development of the number systems of algebra, in addition to those provided by the SMSG texts referred to previously. We list a few of them here:

1. For quite detailed treatments see [H4], [L1], and [L9].
2. For less complete treatments (but still beyond most beginning students) see [E11], [F6], and [H1].
3. For partial treatments suitable for general student reading see [C12], [C22], [D15], [J8], [K7], [R15], and [S39].

3. The Rational Number System

Operations with positive rational numbers fall under the province of arithmetic and will not be discussed here.*   Negative numbers are also familiar to the student from elementary algebra, but it is my experience that a rapid but thorough review of the arithmetic of positive and negative numbers is needed in intermediate algebra to avoid later difficulties.   (See [M14] for an interesting article on the history of the opposition to the use of negative numbers.)

The so-called number line can be used to great advantage here to make intelligible the concept of positive and negative numbers and (with one exception) the rules for operating with positive and negative numbers. The one exception is the rule "minus times a minus is a plus." The reader, of course, is familiar with the definition of integers, in terms of ordered pairs of natural numbers, that makes this "mysterious" rule a consequence of the definition of the multiplication of ordered pairs. This device should certainly be made available via references (e.g., [B25], [E11], [H4], [L9]) to the more able students in college algebra (and perhaps even intermediate), but it is certainly beyond the grasp of the average student (at least without spending a disproportionate amount of time on it).

---

* Not that the teacher should expect perfection in arithmetic from the beginning student. In fact, it is often profitable to spend a little time in intermediate algebra in reviewing the basic facts of arithmetic.

A considerable amount of writing has been done on the establishment of the rule $(-a)(-b) = ab$ in the beginning classes—much of it showing a lamentable lack of understanding. The consensus of informed opinion* seems to be that it is best to present the rule as a definition that proves useful, together with one or more plausibility arguments. For example, in [R15] the author argues thus: Since $2 \cdot 3 = 6$ it is logical to expect $(-2)(-3)$ to be either $+6$ or -6. Suppose it is -6. Then, since $(-2)(+3) = -6$, $(-2)(-3) = (-2)(+3)$, and *if* we want the cancellation law to hold, we must have $-3 = +3$. In [K19], pp. 26–27, the author gets $(a-b)(c-d)$ for $a > b$, $c > d$ by area of a rectangle and then takes $a = c = 0$. He adds, of course, that " . . . the rule of signs is not susceptible of proof; one can only be concerned with recognizing the logical permissibility of the rule." See also [B38] and the references of Section 4.2.

4. The Fundamental Operations

Students rarely have any great difficulty with the addition, subtraction, multiplication, or division of polynomials. They do need drill, of course, especially in division, but even the poorer students eventually do well in this. Parentheses removal, too, is fairly well grasped after drill.

The reasoning behind these operations, of course, is another matter, and an explanation of them ties in well with the axioms (as mentioned before). The instructor should use care, however, that he not make the work too lengthy and boring to the average student by excessive harping on the theory.

There seems to be little point in drill on very complicated problems in the fundamental operations. Problems met with in multiplication outside the algebra course are usually simpler than the more difficult ones found in most algebra books. Division of one polynomial by another is infrequently used, except when the divisor is a monomial or a binomial. In division by a

---

* That is, leaving out those writers who believe that they can prove the rule by heuristic argument!

binomial, synthetic division is a very useful tool, of course, and should certainly be taught. Parentheses occur with extreme frequency throughout mathematics, but there is rarely a need for more than two symbols of grouping.

5. Factoring

This is a difficult topic for many students—the exact degree of difficulty depending upon how extended a job is done on factoring. There is an increasing tendency to eliminate many of the more complicated types of factoring, and a survey of the ways in which factoring occurs (see below) supports this trend.

Perhaps the most important way to assist the student in overcoming his difficulties is to get him to read such identities as

$$a^2 - b^2 = (a + b)(a - b)$$

as "(1st quantity)^2 minus (2nd quantity)^2 equals (1st quantity plus 2nd quantity) times (1st quantity minus 2nd quantity)." In this way he will be better able to handle such problems as the factorization of $(a + b)^2 - c^2$, which frequently baffles students even when they can factor $a^2 - b^2$ easily.

Problems involving the factoring out of common terms, the difference of squares, and trinomials occur in great frequency in mathematics, but the other types usually discussed occur very infrequently. In place of extensive work in intermediate algebra on these little used types, I would strongly urge work on a type of factoring that is used quite frequently in calculus, at least, and that is often not treated in intermediate algebra courses. This is the type where the factors of a polynomial in one variable are sought, and it turns out that one or more of the zeros of the polynomial are rational (and, in fact, usually integral). The rule for determining the possible factors is easily comprehended by students, and the proof of the rule is within the grasp of the better students at least. The actual trials are certainly best performed by synthetic division and the student should be taught to observe the importance of changes in sign of the remainder. Thus, for example, if the remainder in division by $x - a$ is positive, and the remainder in division by $x - b$ (where $b > a$) is negative, there is a factor $x - c$ with $a < c < b$.

In [D12] the authors strongly advocate practice on examples involving subscripts, decimal coefficients, and upper and lower case letters because of the occurrence of these types in applications and general student difficulties with such examples.

Our discussion so far has dealt with the mechanics of factoring. It should also be made clear to the student that the factorability of an algebraic expression depends upon the set of numbers with which we are working. Thus $x^2 - 2$ is not factorable (except for "trivial" factorizations such as $\frac{1}{2}(2x^2 - 4)$) over the rational numbers but is factorable over the real numbers; $x^2 + 1$ is not factorable over the real numbers but is factorable over the complex numbers. It should also be stressed that there exists no set of rules whereby all factorizations of polynomials can be accomplished (cf. remarks in Section 4.12).

6. Word Problems—General Methods

Word problems are commonly considered under several sections, such as those leading to linear equations, those leading to fractional equations, those leading to quadratic equations, etc. However, we shall consider the entire subject of word problems as one unit here. This topic has been more extensively written on than any other topic of algebra. Actually, of course, word problems are not unique to algebra, and much of what is said here has application to other mathematics courses.

There is little question but that word problems are the most difficult part of any mathematics course for most students. On the brighter side of the picture, however, is the fact that, once students begin to enjoy some success in solving word problems, this topic rates high in student interest. In fact, a strong argument can be made for motivating all drill work in manipulation by means of word problems. See, for example, [L12].

The best way to begin the study (actually review in the case of intermediate and college algebra) of word problems is to use "partial" word problems (see [Y6]). For example, "If Frank is twice as old as Henry, and if A represents Henry's age, what represents Frank's age?" Vice versa, there should be training in translating from equations to words as, for example, "If v represents the velocity per second in feet and if s represents the number of feet moved in t seconds, what does $s = vt$ say; what does $v = s/t$ say; $t = s/v$?"

In such drill, as well as in the solution of actual problems, it should be emphasized that, for example, the symbol "$=$" may stand for the following words or sets of words: equals, is, the same as, what is left is, the result is, balances, gives, leaving, makes, etc. Similar remarks can be made concerning the symbols $+$, $-$, $\times$, and $\div$.

After drill in partial problems the student is ready to tackle actual problems. At this point many teachers and authors seem to despair of ever getting the student to understand word problems, and therefore they neatly divide their problems into types (such as mixtures, rates, etc.) and proceed to describe mechanical schemes (sometimes involving carefully drawn "boxes") to solve each type. Anyone who has tried to teach word problems knows the temptation to fall into this defeatist frame of mind. But while such artificial devices may serve the purpose of having students pass examinations that are carefully constructed along the lines of the textbook, they certainly do relatively little to develop the ability to solve the different types of word problems encountered in later mathematics and in the sciences.

Although word problems can never be reduced to a routine, there are two devices which, I believe, can greatly contribute to the student's ability to handle them. The first I call the "cross-out" technique. This procedure helps the student to avoid being overwhelmed by a large collection of words. For example, consider the following simple problem.

From a cylindrical tank, one half full of crude oil, 100 gallons are removed, and 36 gallons are lost by evaporation and leakage. The tank is then one fifth full. How much does it hold when full?

The first step is to cross out all extraneous data. (It is suggested that the student *actually* cross out and not just do it mentally.) The problem will then appear as follows:

From a cylindrical tank, one half full ~~of crude oil,~~ 100 gallons are removed, and 36 gallons are lost ~~by evaporation and leakage.~~ The tank is then one fifth full. How much does it hold when full?

The question sentence is then crossed out and replaced by "x gallons." (To save space we will not write the problem over again each time with the successive cancellations.) We now bring into play the second of our tools—the device of the "basic equation" (called the "key equation" in [P3] and "characteristic formula" in [L12]). The basic equation is an intermediary between the purely verbal given statement of the problem and the final purely algebraic. Here our basic equation is:

Gallons in half full tank $- 136 =$ gallons in one fifth full tank

(at this point crossing out the phrase "100 gallons are removed and 36 gallons are lost"). The last step, of course, is to replace the phrase "gallons in half full tank" by $(\frac{1}{2})x$ and the phrase "gallons in one fifth full tank" by $(\frac{1}{5})x$ and, as we do this, to cross out the respective phrases in the text of the problem.

We give one more example of this technique:

How many gallons of a mixture containing 80% alcohol should be added to 5 gallons of a 20% solution to give a 30% solution?

Here our basic equation is:

Amount of alcohol to start with in 20% mixture $+$ amount of alcohol added $=$ amount in final mixture.

Then the query "How many gallons of a mixture . . . " is replaced by x gallons (thus crossing out this phrase). Continuing, we replace the second term of the basic equation by $0.8x$ (crossing out "contains 80% alcohol" in the problem); "should be added to" is crossed out because of the plus, etc.

All of this "basic equation" method is spelled out in great detail with further embellishments in [L12] (intended for high-school teachers). Many really difficult problems are worked out by this method, and the reader is referred to this book for further details.

7. Word Problems—Further Remarks

The problems to be considered by the class are usually determined largely by the textbook they are using. However, a few remarks concerning the choice of problems are in order. In an excellent book, [S5], the author states that good problems

> . . . should be concrete and vivid. It is more important that they be real to the student now than that they be of practical value to him later. Other things being equal, that problem is best that appeals to the child as being most valuable and important. Similarly, so far as mathematical training is concerned, the one which demands the least explanation of its mathematical content is the most desirable. If our aim is to give practice in computation, we should use data of the sort that accompanies real situations, but if our object is training in analysis, we are justified in altering our data so that the numerical part of the problem may not inhibit the analytic side of the work.*

She also states that students are not amused by or interested in elaborate details invented to buttress a really useless problem. Other suggestions that she gives for improving word problems are:

1. More numerous illustrations of the use of algebra in real situations.
2. More careful choice of problems that appear to be real and important.
3. Greater emphasis on the method of attack, rather than on the answer itself.
4. The use of a larger interest span in discussing solutions.
5. More frequent use of general cases.
6. Fewer problems more carefully considered.

All of these points are worth considering, but I would particularly like to emphasize 5. The general cases are always considered the most difficult by students, and yet they are really the important ones, since, if any type of problem occurs frequently, a formula will be constructed to handle its solution.

* Reprinted by permission of the Bureau of Publications, Teachers College, Columbia University.

Thorndike has said, "It degrades algebra to invoke it to do what the pupil can do better without algebra." This point is elaborated on in [N10]. For example, it is not good to ask the student to use the formula $c = np$ to find the cost of 5 pounds of sugar at 7 cents a pound. *But* we can ask him to write as a formula the rule used to find the cost of 5 pounds of sugar at 7 cents a pound. Along the same lines, there is an amusing article [L11] in which the author presents very forcibly the fact that a great many word problems customarily given in algebra books are capable of easier arithmetic solution (e.g., the courier problems—but many much more difficult problems are considered in this article).

Both the book [L12] previously referred to and the article [L6] criticize the usual delay in the introduction of systems of equations. Both authors advocate an earlier introduction so that the student may decide for himself just how many unknowns he wants to use.

The value of an occasional problem with no solution is pointed out in [D17] and in [S28].

The history of the "classical" word problems in algebra has been well treated in [S5]. See also [S29].

Finally, I claim that the education of any algebra class has been sadly neglected if they have not had read to them Stephen Leacock's hilarious "Human Interest Put Into Mathematics," [L7] (the famous story about A, B, and C).

There is no need, I am sure, to stress the importance of word problems in the sciences. Beginning teachers may not realize, however, how widespread is the complaint of science teachers that students cannot do word problems outside the mathematics classroom. See also [M3], [O2], [R28], and [T6].

8. Algebraic Fractions

Algebraic fractions are no more popular with high school and college students than arithmetic fractions are in grade school. The common difficulties are connected with "cancellation" and with the failure to realize that a negative sign preceding a fraction applies to *each* term of the numerator. For example, we have*

$$\frac{2\cancel{x} + 5}{5\cancel{x} + 3} = \frac{7}{8}, \qquad \frac{x-1}{x+2} - \frac{2x-5}{x+2} = \frac{x-1-2x-5}{x+2}$$

The first is a good example of the famous Freshman Theorem: "Whenever two x's appear on the paper at the same time they may be canceled." An even more elegant example of the use of this "theorem" is afforded by the "proof" that $(a^2 - b^2)/(a-b) = a + b$ as follows:

$$\frac{a\cancel{2} - b\cancel{2}}{\cancel{a} - \cancel{b}} = a + b$$

(since "a minus divided by a minus is a plus"). Still another one is $1\cancel{6}/\cancel{6}4 = 1/4$. (The only fractions with two-digit numerators and denominators with similar properties are—except for the trivial ones such as $22/22$—$16/64$, $19/95$, $26/65$, and $49/98$.)

Because of the misuse of the word "cancellation," both in work with fractions and in work with equations, many teachers have advocated the elimination of the use of the word. Properly understood, however, it is a useful shorthand for lengthy statements, and I question whether it can or should be entirely abolished. It would be well to use it sparingly at first and to ask students from time to time what "cancellation" really means.

Both types of errors mentioned above can be eliminated only by constant pressure. In the first case, continued reference to numerical counterexamples is most helpful. In the second case, we must stress the writing of an intermediate step, as

$$\frac{x-1}{x+2} - \frac{2x-5}{x+2} = \frac{(x-1)-(2x-5)}{x+2}$$

until the students have overcome their tendency to perpetrate this error.

At one time the chapter on fractions in advanced algebra

---

*See [R17] for a fine collection of similar errors of freshmen. The author points out that many teachers are at fault in using a formal procedure which fosters such mistakes on the part of the student.

textbooks ended with very complicated complex fractions.   Nowadays most books confine themselves to the simpler varieties. Although simple algebraic fractions are constantly met with in advanced mathematics and in applications, I have rarely found a complex fraction used of a greater order of difficulty than

$$\frac{\dfrac{a}{b} + \dfrac{c}{d}}{\dfrac{u}{v} + \dfrac{x}{y}}$$

except in the theory of continued fractions (with which very few students will have contact).   Here, of course, the student should learn that the most efficient method of simplification is to multiply both the numerator and the denominator of the complex fraction by the L.C.D. of the simple fractions.   Thus

$$\frac{\left(\dfrac{a}{b} + \dfrac{c}{d}\right) bdvy}{\left(\dfrac{u}{v} + \dfrac{x}{y}\right) bdvy}$$

rather than

$$\frac{\dfrac{ad + bc}{bd}}{\dfrac{uy + vx}{vy}} = \frac{ad + bc}{bd} \div \frac{uy + vx}{vy},$$

etc.

9. Exponents, Radicals, and Irrational Numbers

Positive integral exponents cause little trouble for the student, but as soon as negative and fractional exponents are introduced he begins to flounder badly. As far as the mechanisms of the subject go, no remedy has been suggested beyond that of extensive drill, an insistence on the fact that all work is based on a very few definitions and laws, and an elimination of artificial problems that find no application. On the last point, for example, such monstrosities as

$$\left(\frac{3a^{-\frac{1}{2}}b^{-2}}{a^{\frac{5}{2}}b^{-\frac{1}{2}}c^{3}}\right)^{-\frac{4}{6}}$$

appear only in algebra books. It is far better, in my opinion, to give extensive drill work in simple problems together with a constant emphasis on meaning, than to assign such artificial examples as the above.

In addition to the difficulties of manipulation, there are also complexities of theory. First of all there is the question of the existence of irrational numbers. Certainly at this point it should be proved that $\sqrt{2}$ is irrational.* The usual *reductio ad absurdum* form of proof is probably the best, but the teacher might be interested in seeing four others given in [E10].

It is questionable how far one should go with the average class beyond this point. It is all too easy to get into an involved lecture that is over the head of the class on this topic. Unless the class shows real interest through questions, it is probably best to confine one's remarks to a brief historical résumé of the subject of irrational numbers together with references for the better students, such as [C22], [F2], and [R15]. The first reference gives both the Cantor and the Dedekind treatment in a relatively simple way on pp. 58–72 and a discussion of algebraic and transcendental numbers on pp. 103–109. The second reference has quite a detailed discussion of Dedekind cuts on pp. 208–213, while the third has a good discussion of why it is remarkable that we need irrational numbers, as well as a very careful proof of the irrationality of $\sqrt{2}$, on pp. 104–109. For more detailed treatments see, for example, [K7] or [L1].

In the treatment of radicals, the same admonition to avoid excessively complicated expressions applies. Textbooks, how-

---

*G. H. Hardy in his excellent little book *A Mathematician's Apology* uses the proof of the irrationality of $\sqrt{2}$ as one of two examples of easily understood cases of "real" mathematics (the other is the proof of the infinitude of primes).

ever, frequently mention only rationalization of the denominator and not rationalization of the numerator. Since the latter is used occasionally (as in the differentiation of $\sqrt{x}$ by the Δ-method), it should be brought in if omitted by the text.

Some authors still fail to point out that $\sqrt{x^2} \neq x$ if x is negative. Thus they obtain $\sqrt{x^2 + 2x + 1} - \sqrt{x^2 - 2x + 1} = \sqrt{(x + 1)^2} - \sqrt{(x - 1)^2} = (x + 1) - (x - 1) = 2$. But if $x = 0$, $\sqrt{x^2 + 2x + 1} - \sqrt{x^2 - 2x + 1} = \sqrt{1} - \sqrt{1} = 0$. The correct statement, of course, is $\sqrt{x^2} = |x|$.

Further references are [L3], [L4], [R11], [S38]. In [M16], the author analyzes the algebra topics in which students of calculus and other advanced topics have weakness. In particular, he concludes that more stress needs to be placed on fractions, exponents, and the addition of radicals.

10. Complex Numbers

There are actually few student difficulties in the mechanics of complex numbers. The trouble lies, rather, in what Gauss called "the true metaphysics of complex numbers." Too often it seems to the student that the textbook or instructor (or both) first argue strongly that there is no such thing as the square root of -1, and then turn around, call it i, and proceed to use it.

The true metaphysics lies, of course, in the representation of complex numbers as ordered pairs of real numbers, and it seems feasible to me to attempt such a treatment to a capable advanced algebra class. An excellent reference for this purpose is [D15] pp. 72–96, and the SMSG text [S16]. Incidentally, in [D15] the author suggests calling numbers of the form *"ib" normal* numbers, since, in the vector representation, these vectors are perpendicular to those representing real numbers (see also [S11]).

For intermediate-level classes, however, such a treatment is probably too difficult, and the vector approach is suggested here. In such an approach we can think of the quantity i as a "rotation factor," which rotates unit vectors through 90°. Then, applying the rotation factor twice, we get the negative unit vector and thus have a heuristic argument to the effect that $i \times i = -1$ (e.g., see [F2], Chapter 13).

Complex numbers are, of course, used extensively in many applications of mathematics, as well as in advanced pure mathematics. It is also possible to give more immediate applications of complex numbers in the proof of theorems in geometry. See, for example, [B23], [S10], [S23], and [S27].

11. Solution of Equations in One Unknown—Elementary

The solution of linear equations in one unknown* presents little difficulty to the student except when literal coefficients are involved (e.g., $ax + c = bx + d$, $a \neq b$). These general types should certainly be stressed, both in the linear case and elsewhere because of their frequent occurrence in advanced mathematics, and in applications. Considerable justified criticism has been made of the use of the word "transpose" (cf. remarks on "cancellation"), and it would be well to avoid the use of this word, at least at first and, if it is used eventually, to ask students from time to time what the reasoning is behind "transposition." Note also that in the proof of the fact that $x = -b/a$ is a root of $ax + b = 0 (a \neq 0)$, authors frequently show only that *if* the equation has a root it must be $-b/a$. To show that $-b/a$ is a root involves observing that $a(-b/a) + b = 0$.

Fractional equations present almost the same difficulties as do operations with fractions. The necessity for checking answers to make sure that no denominator takes on the value zero should be stressed, as in $x + 1/(x - 2) = 2 + 1/(x - 2)$. (Incidentally,

---

*The use of the word "unknown" has been justly criticized. In the equation "$x = 2$" we certainly know that $x = 2$, and there is nothing unknown about it. Even in an equation such as $x^2 - 5x + 6 = 0$, where the solutions are (slightly) less obvious, we eventually do learn that $x = 2$ or $x = 3$. Furthermore, in an equation such as $x - a = 0$, we would normally call x the unknown, and yet when we solve for x, we obtain $x = a$, where, in some sense, a is unknown. We may, of course, use an alternate word such as pronumeral or write, for example, $\{x : x^2 - 5x + 6 = 0\} = \{2,3\}$. The use of the word unknown, however, is traditional and is likely to continue in spite of logical objections to it.

both here and in the solution of equations involving radicals, the term "extraneous roots" is often used. I believe it is confusing to call a number a root [extraneous or otherwise] when it is not!) A frequent phenomenon that occurs after students have had work with the elimination of denominators in fractional equations is that, when they return to the addition of fractions, they often omit the denominator in the answer! Watch for this.

At this point I want to mention [B37], which deals with the important question of the equivalence of equations in one unknown. The author shows that the usual rules given as permissible ones for the transforming of equations may lead to difficulties. For example, the equation $x^2 - 5x + 6 = 0$ has the roots 2 and 3. Adding the same quantity, $1/(x-2)(x-3)$, to both sides we have

$$x^2 - 5x + 6 + \frac{1}{(x-2)(x-3)} = \frac{1}{(x-2)(x-3)},$$

which is an equation without roots. Further examples are given to show that "doing the same thing to both sides" doesn't always work out well. Finally the author gives correct rules as follows:

1. The addition of any constant or any polynomial in the unknown to both sides of an equation of condition gives an equation equivalent to the given equation. Similarly for subtraction.
2. The multiplication of both sides of an equation of condition by any constant save zero gives an equation equivalent to the given equation. Similarly for division.
3. Given an equation of condition consisting of fractions whose numerators and denominators are constants or rational integral functions of the unknown; if (1) the fractions are all in lowest terms, and (2) the denominators are relatively prime; then the result of clearing of fractions gives an equation equivalent to the given equation.

The proof of 1 and 2 can be found in [C8], Chapter XIV, and of 3 in [Y5], Chapter 5.

One other general point: It is distressing to find students who think, for example, that $x^2 - 5x + 6 = (x - 2)(x - 3)$ is not an equation or, if they admit that it is an equation, feel it has no solutions.

The solution of quadratic equations either by factoring, completion of the square, or by formula usually affords little difficulty if enough drill is provided. The method of completion of the square is the most difficult, and I find it helps to present six examples at one time, ranging from the simplest to the general case. That is, I list the six equations at the top of six sections of the board, then do step 1 for all six, then step 2, etc. This method of presentation involves quite a bit of walking, but it does put it across! In solution by formula, students frequently apply the Freshman Theorem to obtain, for example,

$$\frac{2 + \sqrt{5}}{2} = 1 + \sqrt{5}.$$

It should be noted that while quadratics are infrequently solved in practice by the completion of the square method,* completion of the square is frequently used in analytic geometry and in calculus. Thus, for example, one writes $x^2 - 2x + y^2 - 6y + 6 = 0$ as $(x - 1)^2 + (y - 3)^2 = 4$ to find the center and radius of the circle with equation $x^2 - 2x + y^2 - 6y + 6 = 0$; and one writes $x^2 - 4x$ as $(x - 2)^2 - 4$ in connection with the integration of $\sqrt{x^2 - 4x}$. I strongly recommend the inclusion of problems like those encountered in these subjects at this point.

It will also interest students to see at least one other derivation of the quadratic formula. The one that uses the substitution $x = y + m$ to eliminate the linear term is particularly recommended, as the same device is used in the solution of cubics and quartics. Thus we consider $ax^2 + bx + c = 0$, replace x by $y + m$, and find that when $m = -b/2a$, we have $y^2 = (b^2 - 4ac)/4a^2$. Hence $y = \pm \sqrt{b^2 - 4ac}/2a$, and $x = y + m$ (see [D18], p. 32). Other methods are given in [C11], [F3], and [P12]. The better students might be asked to criticize the second method of derivation given in the last reference. Here the author lets

---

* However, some engineers seem to prefer this method. See, for example, [W12].

$x = m + ni$ in $ax^2 + bx + c = 0$ and then equates to zero the coefficients of the real and imaginary parts, thus tacitly assuming that a, b, and c are real numbers.

Discriminants of quadratic equations are useful in analyzing equations of conics in analytic geometry. Such problems as "Find the restrictions, if any, on x in order that y be a real number if $y^2 - 2xy - 2y - 4x^2 - 8x - 4 = 0$" should certainly be included for this reason. For maximum usefulness in further work, however, discriminants of quadratic equations should be presented in a way that generalizes to higher-degree equations. See, for example, [A3], [B25], and [F2]. This is a good place, too, to remind students of the fact that the conditions of a theorem must be met before the consequences can be drawn, by presenting "contradictions" of the theorems on the discriminant by ignoring the hypotheses concerning the coefficients of the quadratic equation. Thus, for example, the equation $ix^2 + 3x + i = 0$ has discriminant $9 - 4i^2 = 13 > 0$, and yet its roots are not real numbers. No contradiction, however, is involved since the theorem that states that the roots are real if the discriminant is positive has as hypothesis that the coefficients be real numbers.

12. Solution of Equations in One Unknown—Advanced

Intermediate algebra usually goes no farther than the quadratic equation, but it would be a grave mistake, to my mind, if the instructor did not sketch out for the class the further development of the subject. This subject of the solution of equations presents, in miniature, much of the entire history of mathematics and is most instructive. To learn that quadratics were solved by the Babylonians over 4,000 years ago gives the students historical perspective, and to learn of the tragic stories of Abel and Galois is a startling revelation to students of almost the same age as were those geniuses when they died. Furthermore, it gives us an opportunity to stress how new numbers came to be invented in order to meet the demands for solutions of equations and to point the tragicomic story of the Tartaglia-Cardan imbroglio. See, for example, [B11], [B13], [D13], and [F5].

The treatment of equations of higher degree in college algebra merges almost imperceptibly into the material often taught in a course in the theory of equations. It is generally agreed today that there is no need for this separate course in the theory of equations, although the useful computational aspects of the subject should certainly be taught in some course. Thus a good course in advanced algebra can contain much of these useful computational aspects and can be followed by a course in abstract algebra which treats additional traditional topics in a more modern setting. See, for example, the CUPM recommendations [C16] and also [W11].

13. Systems of Linear Equations and Determinants

In intermediate algebra, attention is usually confined to a systematic treatment of two equations in two unknowns and three equations in three unknowns, plus a little work on four equations in four unknowns. Usually the methods of solution by addition and subtraction, substitution, and determinants are given. Little trouble is experienced by students with this topic, and we have already noted that writers have urged that the choice of the number of unknowns in the solution of word problems be left up to the student at the very beginning, and this implies an earlier treatment of this topic than is usually given.

The customary diagonal method of evaluating third-order determinants has been deservedly criticized, however. It is just as easy to evaluate them by expansion by minors, and this method, of course, can be generalized, whereas the other cannot. When the technique is introduced of adding multiples of one row (column) of a determinant to another row (column), students frequently feel that there is great virtue in always obtaining a determinant in which *all* of the elements are zero except those on the diagonal. In practice, this is almost always a waste of time, and the student should be instructed to evaluate a determinant by whatever method involves the least numerical computation.

The general case of n equations in n unknowns, including a treatment of determinants of order n, is often taken up in college algebra, although it seems reasonable to me to include this in a course in modern algebra along with matrices. (For an elementary derivation of Cramer's rule see [W15].) In any case, no matter how much or how little of determinants is discussed, the double-subscript notation should certainly be introduced, as it is used so frequently elsewhere.

The classical reference for determinants is, of course, the four volumes of T. Muir [M22]. There is also an excellent article [P15] that includes a history and a bibliography and in which modern matrix methods of proof are given. See also [R25] where the following theorem is proved.

Suppose given a system of linear equations (1) $\sum_{i=1}^{n} a_{ki}x_i = b_k$ ($k = 1, 2,$ $\ldots, n$). If it is possible to obtain the equations (2) $x_1 = c_1$, $x_2 = c_2$, $\ldots, x_n = c_n$, by forming linear combinations of the given equations, then (2) is a solution of (1) and is, of course, the only solution.

The proof involves matrix theory and affords an excellent example of the use of advanced mathematics to shed light on a very elementary process. (It is quite feasible and very much in order, by the way, to introduce students in an intermediate algebra course to at least 2×2 matrices.) For example see [D18].

14. Systems of Quadratic Equations

This is another topic which, in my opinion, can easily be over-emphasized. Nearly all the applications I have seen are those in which one of the equations is linear, with just a scattering of the following types:

(1) $Ax^2 + By^2 + Cx + Dy = E$, $Ax^2 + By^2 + C'x + D'y = E'$;
(2) $xy = c$ and the second equation a general quadratic;
(3) $ax^2 + by^2 = c$, $a'x^2 + b'y^2 = c'$.

For the elementary theory see [K4] where the author gives conditions under which the solution of a pair of simultaneous quadratic equations will reduce to the solution of quadratics. The general theory is, in my opinion, only of historical interest today.

15. Progressions

This is a relatively simple topic that most students enjoy, and it might well be considered earlier in intermediate algebra to provide some relief from the emphasis on manipulation in the usual first few chapters.

This is a good place, it seems to me, to give the widely used subscript notation a good workout instead of writing $a, \ldots, l$. Along the same line, it is better to speak of the nth term instead of the last term and to emphasize that we can relate the concept of a sequence to the concept of a function. Thus we may define a sequence function as one with domain the set of natural numbers and point out that $s_1, s_2, \ldots, s_n$ is simply an alternative way of writing $s(1), s(2), \ldots, s(n)$.

The artificial nature of the usual proofs for the sum of the arithmetic and geometric series (especially the latter) has often been condemned. An alternative more general approach is considered under induction.

Although the evaluation of repeating decimals as fractions affords an excellent example of the use of geometric series, it seems only fair to point out to the student that an easier and less sophisticated method exists. For example, if $x = 0.1111 \ldots$, $10x = 1.1111 \ldots = 1 + x$, $9x = 1$, $x = 1/9$. (A similar technique is often useful elsewhere. Thus if we wish to evaluate

$$x = \sqrt{2 + \sqrt{2 + \sqrt{2 + \sqrt{2}} \cdots}}$$

we have $x^2 = 2 + x$, so that $x = 2$. In particular, continued fractions such as

$$\cfrac{1}{1 + \cfrac{1}{1 + \cfrac{1}{1 + 1} \cdots} \cdots} \cdots$$

can be evaluated in this fashion.)

For further work see [S40] in which the series $\sum\limits_{k=1}^{\infty} k^n x^k$, where n is a non-negative integer and $|x| < 1$, is considered.

Both arithmetic and geometric progressions find occasional use in such subjects as calculus, mechanics, optics, and kinetic theory, and, of course, geometric series lie at the very heart of the mathematics of finance.

16. Induction

This important tool for all of mathematics is a difficult one to sell. Students find it hard to grasp the nature of proofs by mathematical induction and, I feel, regard such proofs as being slightly illegitimate. On a very elementary level there is an article [W9] on a way of presentation to remove this doubt, in which the author presents the basic ideas in the following form:

> Kuedee wants to get people in a line at one theater to switch over to his. He asks Kayplusone who says he will go if Kay does and adds that his attitude is typical. Then Kuedee goes to One, etc.

Faced with the difficulty of teaching this topic, many textbook writers and teachers make a mechanical process out of it. That is, given a certain conjectured equality involving integers, they teach students to go through a routine of checking for $n = 1$, writing the conjectured equality with n replaced by k, changing k to $k + 1$, performing manipulations, etc. It seems to me that if it is worthwhile to present induction at all, it is better to emphasize its more general character in the form: If a proposition, $P(n)$, is defined for every positive integer n, and if $P(1)$ is true and if the truth of $P(k)$ implies the truth of $P(k + 1)$, then $P(n)$ is true for all n.* It should be emphasized that this is an axiom and not something which is a "self-evident truth."

Still another formulation uses the idea of an inductive set.

---

* The better students might even be exposed to the equivalence of the principle of induction to the well-ordering principle. See, for example, [B25], pp. 9–10.

This is extremely well presented in a film produced by the Mathematical Association of America, with an accompanying pamphlet [H20] written by the "star," Leon Henkin. See also the article by Henkin [H20a].

Along with a more general formulation should go some uses of mathematical induction other than the traditional ones of proving that a certain sum of n numbers is expressible as a polynomial in n. (As, e.g., that $a^n a^m = a^{n+m}$ for m and n positive integers where $a^1 = a$, $a^{k+1} = a^k a$ for $k \geq 1$, or that $n^3 + 1 \geq n^2 + n$). This more general procedure is followed in [C22], pp. 9–20, and in [R15], pp. 392–401. At the conclusion of the former reference, the authors have a very interesting "proof" by mathematical induction that any two positive integers are equal as follows: Let A_n be the statement, "If a and b are any two positive integers such that max $(a,b) = n$, then $a = b$. Then (1) A_1 is true for max $(a,b) = 1$, $a = b = 1$; (2) If A_r is true, let a and b be any two positive integers such that max $(a,b) = r + 1$. Consider $\alpha = a - 1$, $\beta = b - 1$. Then max $(\alpha,\beta) = r$. Hence $\alpha = \beta$ since we are assuming A_r to be true. Hence $a = b$, A_{r+1} is true." (The hitch, of course, is that b might be 1 in (2), and then $\beta = 0$ is not a positive integer.) A more sophisticated example of essentially the same "theorem" involving the Peano postulates is given in [A4]. See also [E11].

Returning to the subject of finite sums,* we find that many authors ask the student to write the general term of, for example, the sequence 2, 4, 6, 8, ... Such a request calls for clairvoyance since the nth term in the example can be written as $2n + (n - 1)(n - 2)(n - 3)(n - 4)f(n)$, where f is any function of n whatsoever. In these exercises, too, the students often wonder how the formulas on the right are determined. In [M19] the author suggests, for example, setting

$$1 + 3 + \cdots + (2n - 1) = a + bn + cn^2$$

and letting $n = 1$, 2, and 3 to get a, b, and c. In general, if the generating term of a series to be summed to n terms is a polynomial in n of degree k, the sum is a polynomial of degree $k + 1$

---

*The very useful summation symbol Σ should certainly be stressed here as well as elsewhere.

in n. While the theorem can easily be proved by the calculus of finite differences, it can be used without proof. Thus obtaining $a = b = 0$ and $c = 1$ in $1 + 3 + \cdots + (2n - 1) = a + bn + cn^2$ can be interpreted as showing that *if* there are numbers a, b, and c, such that $1 + 3 + \cdots + (2n - 1) = a + bn + cn^2$, they must have the values $a = b = 0$ and $c = 1$. The proof that these values are indeed correct is then made by induction. Thus it is first shown that it is *necessary* to have $a = b = 0$ and $c = 1$, and then shown that it is sufficient to have $a = b = 0$ and $c = 1$. (Frequent discussions of the distinction between necessary and sufficient conditions, the meaning of "if and only if," etc. are *very* important to the students' success in mathematics.)

Students always enjoy the "proof" that all positive integers are interesting. For 1 is certainly interesting as being the first positive integer, the only positive integer whose square is itself, etc. Assume now that k is the last interesting positive integer. Then $K + 1$ is the first noninteresting integer; but this is itself an intensely interesting fact about $k + 1$.

For a historical treatment see [B43]. Other articles on induction are [B12], [H24], and [R6].

17. The Binomial Theorem

Students generally enjoy this topic and soon learn to be adept at the expansions. All that is needed is sufficient practice. Particular stress should be laid on the use of the binomial theorem for approximations, especially with negative and fractional exponents, as it is in this way that it is most frequently used in the sciences.

Some references are [C19] (requires a knowledge of calculus in some parts), [S42], [R23], and [T8] (the latter two consider a Pascal triangle for negative exponents). Incidentally [B13] calls attention to the fact that Cardan anticipated Pascal with respect to both the form and the study of the triangle. On the Pascal triangle see also [A7] and [M9]. The latter contains some Pascal triangles for expansion of powers of various polynomials with more than two terms. In [F11] there is a proof of the binomial theorem that does not involve the use of induction.

18. Miscellaneous

Logarithms properly belong in analysis, are often considered in algebra, and are used most extensively for computation in trigonometry. Since the algebra chapter is already quite long we shift this topic to the next chapter.

In regard to ratio, proportion, and variation I note first that these topics are used extensively in physics, engineering, etc. The treatment given in most textbooks is usually adequate except that too much stress is usually placed in finding k (the constant of proportionality). Thus, for example, if a student is told that the volume of a gas varies inversely as the pressure he should not have to find k to answer the question, "If the pressure on a gas is doubled, what happens to the volume?" I have one reference [R27] where a proof of the theorem, "If z varies as x, and z varies as y, then $z = kxy$," is given.

Little use is made of elaborate schemes for finding the G.C.D. or L.C.M. of polynomials. When these expressions are needed they can most often be found by inspection or by simple factoring.

Almost all the more recent algebra textbooks and many of the grade school textbooks consider inequalities along with equalities. Certainly inequalities are important, and even if the book being used does not consider them, it is easy and worthwhile to introduce at least the basic ideas. See, for example, the SMSG textbooks [S14] and [S16], and also [T9].

One of the distinguishing features of a majority of the newer texts in mathematics at the elementary, secondary, and college levels is the systematic use of the language of sets (sometimes, in my opinion, an excessive use). Certainly the use of such language can help to clarify many concepts, and it provides a neat symbolism: for example, the concept of the solution set of an equation as in $\{x : x^2 - 5x + 6 = 0\} = \{2,3\}$; the description of a line as $\{(x,y) : Ax + By + C = 0\}$; the development of rational numbers as sets of ordered pairs, etc. Even if the text being used does not use set notation it is easy and useful to do so, at least in moderation as in the SMSG texts [S13], [S14], and

[S16]. For an illustration of the systematic and elegant use of set language in elementary mathematics see [K7].

Certainly, too, it should be recognized that, regardless of the text being used, there should be careful consideration given to the concept of a function, either as a mapping or as a set of ordered pairs, sometime prior to the study of the calculus, and a distinction should be made between a function and a functional value (f and $f(x)$). See, for example, the SMSG textbooks [S13] and [S16] and also [K7] and [W22].

Students usually greatly enjoy permutations, combinations, and probability (the latter is a good illustration of the profitable use of set language). Little use is made of these topics, however, at the college-algebra level. That is, when probability is used (and, of course, it has a tremendous use) it is almost always at a much higher level than that of the usual college-algebra treatment.

In concluding this discussion of algebra, I mention one other topic that is frequently given insufficient emphasis in courses in intermediate algebra. This is the topic of absolute value. Every opportunity should be given for the student to work with this concept including graphing (e.g., $y = |x - 1|$) and solution of inequalities (e.g., $|x| + |y| < 1$).

Additional references for Chapter 4 are [B9], [N1], [R21], and [W18].

5

The Teaching
of Trigonometry
and Logarithms

Generally speaking, in trigonometry a great deal of time is wasted on too much numerical computation. Few people find occasion to solve the more complicated cases of oblique triangles, and those who do generally learn to do so over again in specialized courses (e.g., surveying). And when oblique triangles are solved they are usually solved by either the sine law or the cosine law, very little use being made of the half-angle formulas or the law of tangents. The analytic side of trigonometry is, of course, of considerable importance, but even here much work; for example, the proving of very complicated identities, is of little value (except as a review of algebraic techniques). See, in this connection [E4] and [H16].

1. Definition of the Trigonometric Functions

Mathematicians generally agree that too much stress has been laid on the trigonometric functions as functions of *angles*.

Both [N1] and [V1] suggest measuring x on a unit circle and defining cos x and sin x by coordinates. From this definition it is possible to derive all the standard formulas of trigonometry without mentioning angles. [C7] gives eleven different ways in which trigonometric functions can be defined. Not included in this list is the method that uses the so-called winding function, as in [D19], or the method of [R18] in which the vector OP is represented by $F(x)$, cos x is the real part, sin x is the imaginary part, and formulas are derived based on $D_x F(x) = iF(x)$.

While everyone is in agreement that the real number aspect of trigonometry should be stressed, there is not general agreement as to whether a course in trigonometry should *begin* with the real number approach. ([B8] presents arguments for beginning with acute angles.)

Whatever method is used, I would certainly use radian measure along with degree measurement instead of deferring the use of radian measure until late in the course. The usual definition of a radian as "the angle subtended at the center of a circle by an arc equal in length to the radius" has been criticized in [B20], since the meaning of "an arc equal in length to the radius" really involves a rather sophisticated application of the limit process. The author's suggestion is to define the angle which is the $(1/\pi)$th part of a straight angle as a radian. Here π is thought of as a number (defined, say, by a series) rather than the ratio of circumference to diameter.

Another serious objection to the definition quoted is that it confuses an angle (a geometric object that is the union of two half lines with a common endpoint) with the *measure* of an angle (a number). Thus we have an angle A symbolized by $\angle A$ and (say), $m\angle A = 30$, where $m\angle A$ means the measure of $\angle A$ in degrees (or $m\angle A = \pi/6$ if we are using radian measure). See, for example, the SMSG textbook [S15].

There is one more minor point concerning the definition of trigonometric functions which deserves mention here. There are some textbooks that do not state as they should that tan $90°$, cot $90°$, etc., are not defined but, instead, have tan $90° = \infty$, etc. While, later on, students will certainly use the word infinity, it is generally agreed that they should not do so until the full meaning of the phrase is explained in, say, the calculus. For students who have a fondness for the word, the following limerick is appropriate:

> There was a professor of Trinity,
> Who found the square root of infinity.
> But in counting the digits,
> He was seized with the fidgets,
> Dropped mathematics and took to Divinity.

2. Solution of Triangles

I pass by the solution of right triangles as this is never a difficult part of trigonometry and mention first an interesting suggestion for the reform of the trigonometry of oblique triangles (as well as other parts of trigonometry). This is given in [A2]. The author suggests first deriving the formula $\cot A = \dfrac{b}{a \sin C} - \cot C$ (see also [F4]). This, together with the law of sines, will solve all cases in oblique triangle theory. It may also be used with the law of sines to yield formulas for $\sin (A + C)$ and $\cos (A + C)$ by replacing b/a in the equation by $\sin B/\sin A$, with the extension to the general angle carried out, as usual, analytically.

The law of sines needs little comment. Its use in the ambiguous case is the only point of difficulty. In addition to being used extensively in the numerical solution of triangles, as in surveying, it is used analytically in many applied fields. [B35] points out that it is necessary to prove the law of sines for right triangles as well as for the two cases considered by most textbooks.

The law of cosines is also used to a moderate extent in the solution of triangles, and to a considerable extent analytically. It is often used to find the resultant of two forces, but the component method is often preferable; and when more than two forces are involved, the latter is definitely superior. While it is generally considered inferior to the law of tangents for computational purposes, [M7] claims that when *proper arrangement is made for the work*, the law of cosines is on a par with the law of tangents, and is superior if only the third side is wanted. (Actually, much of the discussions concerning the most effective way of solving numerical problems have become obsolete in the face of the widespread use of desk calculators for even occasional computations and the use of computers for types of computation that occur frequently). It has also been pointed out in [O5] that we may use the law of cosines to analyze the ambiguous case. For from $c^2 = a^2 + b^2 - 2ab \cos C$ we obtain, by taking in turn, C as an angle of degree measure 0 and 180, $c = a - b$ and $c = a + b$ respectively. Thus $a - b < c < a + b$ by a familiar geometric theorem. Upon solving for c we can make an analysis of the ambiguous case.

I have already given my views on the law of tangents. It is interesting to note that this almost useless law occupies more space in the literature than all other aspects of elementary trigonometry put together. I shall not give references here. Those who are interested can consult the index of the *American Mathematical Monthly* [A6]. There are also many articles on this topic in *School Science and Mathematics*. Similar remarks apply to the half-angle formulas although the literature is not quite as extensive.

In summary, I would recommend that the law of sines and the law of cosines be treated rather thoroughly and that the law of tangents and the half-angle formulas simply be mentioned. In all computational work, stress should be laid on approximations and significant figures. If work on the law of tangents and the half-angle formulas is omitted, there is ample time for work on the use of slide rules. In connection with significant figures in trigonometry, see [N11]. The author states that a compromise* between exactness and simplicity leads to

1. If the sides are measured to three significant figures, the angles should be measured to the nearest ten minutes;
2. If the sides are measured to four significant figures, the angles should be measured to the nearest minute, and work should be done with four-place tables;

---

*See, however, a recent article by Carl N. Shuster in the *Mathematics Teacher*, vol. LV (1962), pp. 649–650, that is highly critical of this rule-of-thumb.

3. If the sides are measured to five significant figures, angles should be measured to the nearest tenth of a minute, and work should be done with five-place tables.

For a general discussion of significant figures see [B1]. This book gives a rather complete treatment of approximations as related to elementary (pre-calculus) mathematics. See also [G5].

3. Identities of a Single Angle*

Under this heading we consider three different types:

1. the reduction identities,
2. the identities stemming from and including what are usually called the fundamental identities, and
3. the problem of finding all of the trigonometric functional values of an angle, when one is given.

The first topic needs careful presentation if the students are to grasp it successfully. I prefer to teach this topic in two sections, the first dealing with numerical computations and the second with identities. In the first section everything is based on the identity

$$\text{T.F.}\ \theta = \pm \text{T.F. (reference angle of } \theta)$$

where T.F. stands for any trigonometric function, reference angle is defined in an obvious fashion, and the sign is chosen by inspection of a diagram.

On the other hand, to remember the reduction *identities*, I write (using $\bar{\theta}$ to indicate the degree measure of the angle θ)

$$\text{T.F.}\ \theta = \pm \text{T.F.}\ (n \cdot 90 \pm \bar{\theta}), \qquad n \text{ an even integer,}$$

$$\text{T.F.}\ \bar{\theta} = \pm \text{co-T.F.}\ (n \cdot 90 \pm \bar{\theta}), \qquad n \text{ an odd integer.}$$

The sign is determined by the inspection of a diagram in which we take $\bar{\theta} = 45$ (or take θ to be any other acute angle). For

*Or number, depending on whether or not we are considering the trigonometry of angles or the trigonometry of numbers.

example, $\cos (180 + \bar{\theta}) = \cos (2 \cdot 90 + \theta) = \pm \cos \theta$, $\sin (270 + \bar{\theta}) = \sin (3 \cdot 90 + \bar{\theta}) = \pm \cos \theta$ with the negative sign chosen in each case because of the following diagrams:

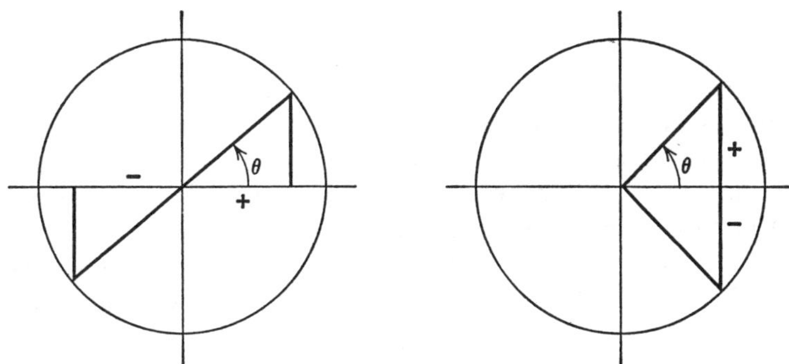

This mnemonic device makes the whole mass of reduction identities easily available. It is not necessary to prove all these identities in full generality. A sample or so will suffice, with others left as student exercises. See also [D9], [D19], and [M15].

The so-called fundamental identities* in all their variations should be stressed. That is, for example, the student should not only be able to recognize instantly that $\sin^2 x + \cos^2 x$ is equal to 1 but also that $1 - \cos^2 x$ is equal to $\sin^2 x$. Now, while we are frequently required to change the form of a trigonometric expression in connection with, for example, integration, there are few places in advanced mathematics or in applications of mathematics where we need to prove an already existing identity. Rather, we are asked to transform a given trigonometric expression into another "more useful" one. The exact nature of "more useful" will, naturally, depend on the context in which the problem is found. So I would recommend less formal work on the proving of identities, and more work on transforming a given expression into another according to given instructions (as, for example, "express in terms of sines and cosines").

---

* That is, $\sin^2 x + \cos^2 x = 1$, $\tan x = \sin x/\cos x$, $\cot x = \cos x/\sin x$, $\tan x \cot x = 1$, $\sec x \cos x = 1$, $\sin x \csc x = 1$, $\sec^2 x - 1 = \tan^2 x$, and $\csc^2 x - 1 = \cot^2 x$.

Closely allied to the use of trigonometric identities is the finding of the other trigonometric functional values of an angle when one functional value is given, since this problem can be done by reference to the fundamental identities. It is better, however, in my opinion, to do this by the drawing of a right triangle, with the signs determined by the quadrant. Be sure to emphasize the general cases (e.g., $\sin x = a$), as these types occur frequently in, for example, integration by trigonometric substitution.

I have already indicated that the reduction identities are used frequently. The fundamental identities occur very often, and, as mentioned before, we often have to transform a given trigonometric expression into another, without the form of the new expression being definitely specified. The third type of problem is used in the calculus (integration by trigonometric substitution) but does not seem to have any other wide application.

For a proof of De Moivre's theorem, by a different method than the usual one, see [L8].

4. Identities of More Than One Angle

There is universal condemnation of the usual way of deriving the identity for $\sin (A + B)$ by the use of a complicated, unmotivated geometrical diagram. Several alternative proofs have been proposed. Among the more interesting of these are those given in [M11], [R29], and [H28].

The first of these proofs requires a knowledge of the distance formula, of the general definitions of the trigonometric functions, and of the identities $\cos^2 x + \sin^2 x = 1$, $\cos 0° = \sin 90° = 1$, $\cos 90° = \sin 0° = 0$. With a vertex O and the half line OW as a beginning, we choose points P and Q on the terminal sides of the given angles A and B respectively, each at a distance of 1 from O. Then with OW as an x-axis we have $\overline{PQ}^2 = d^2 = (\cos A - \cos B)^2 + (\sin A - \sin B)^2 = 2 - 2(\cos A \cos B + \sin A \sin B)$. Then with OQ as positive x-axis, $d^2 = [1 - \cos (A - B)]^2 + \sin^2 (A - B) = 2 - 2 \cos (A - B)$. Equating the two values of d^2, we obtain $\cos (A - B) = \cos A \cos B + \sin A \sin B$. The derivation of the identities for $\cos (A + B)$ and sin $(A \pm B)$ follows quickly by use of the conditions cos $0° =$ sin $90° = 1$, cos $90° = $ sin $0° = 0$. Thus, for example, we put $\bar{A} = 0$ and have cos $(-B) =$ cos B, and if $\bar{A} = 90$, cos $(90 - \bar{B})$ $=$ sin $\bar{B}$. Then if $\bar{B} = 90 - C$, sin $(90 - \bar{C}) =$ cos $\bar{C}$. Thus

$$\sin (A + B) = \cos [90 - (\bar{A} + \bar{B})]$$

$$= \cos [(90 - \bar{A}) - \bar{B}]$$

$$= \cos (90 - \bar{A}) \cos \bar{B} + \sin (90 - \bar{A}) \sin \bar{B}$$

$$= \sin A \cos B + \cos A \sin B.$$

The second derivation uses the identities $a = b \cos C + c \cos B$, $b = c \cos A + a \cos C$, and $c = a \cos B + b \cos A$ to derive identities for sin $(A \pm B)$, cos $(A \pm B)$, tan $(A \pm B)$, cos $C/2$, and sin $C/2$. The third author draws a circle of unit diameter and reads the identity for sin $(A + B)$ from the figure as shown below.

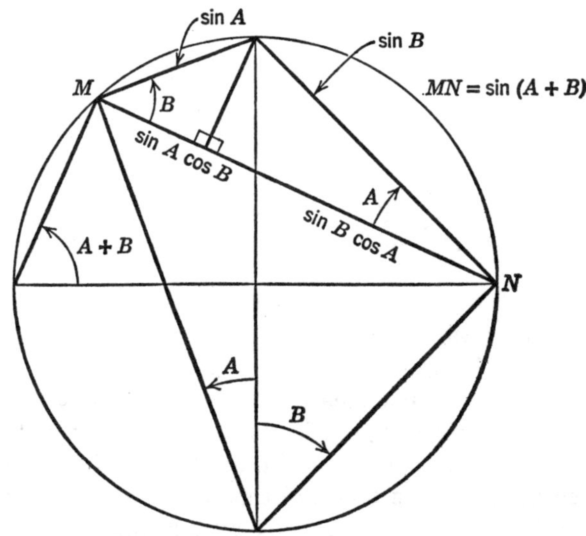

Another approach is via complex numbers. This affords an excellent review and application of complex numbers and provides a natural, easily remembered proof (e.g., see [R18] and [S11]).

Thus if $\alpha_1 = \cos \theta_1 + i \sin \theta_1$ and $\alpha_2 = \cos \theta_2 + i \sin \theta_2$, $\alpha_1 \alpha_2$

$= \cos(\theta_1 + \theta_2) + i\sin(\theta_1 + \theta_2) = (\cos\theta_1 \cos\theta_2 - \sin\theta_1\sin\theta_2) + i(\sin\theta_1\cos\theta_2 + \cos\theta_1\sin\theta_2)$. Then, by equating real and imaginary parts, we obtain the desired identities for $\sin(\theta_1 + \theta_2)$ and $\cos(\theta_1 + \theta_2)$.

Still another approach uses the distance formula to derive the identity $\cos(x+r)\cos(y+r) + \sin(x+r)\sin(y+r) = \cos x \cos y + \sin x \sin y$. From this the usual identities can be obtained. Thus we get an identity for $\cos(y-x)$ by taking $r = -x$. See [D19] for details.

From the addition identities for the sine, of course, the other addition identities, as well as the double and half-angle identities and the identities for the sum and difference of trigonometric functional values, can be obtained analytically. In the usual derivation, however, of $\tan\frac{1}{2}\theta$ by division of $\sin\frac{1}{2}\theta$ by $\cos\frac{1}{2}\theta$, there arises a seeming ambiguity of sign which must be argued away. [W23] avoids this by taking $\sin\frac{1}{2}\theta = \sin(\theta - \frac{1}{2}\theta) = \sin\theta\cos\frac{1}{2}\theta - \cos\theta\sin\frac{1}{2}\theta$. Thus $(1 + \cos\theta)\sin\frac{1}{2}\theta = \sin\theta\cos\frac{1}{2}\theta$, so that

$$\tan\frac{1}{2}\theta = \frac{\sin\frac{1}{2}\theta}{\cos\frac{1}{2}\theta} = \frac{\sin\theta}{1 + \cos\theta}$$

directly.

These multiple-angle identities are all used extensively in many parts of advanced mathematics and in applications, and they should receive considerable attention.

5. Inverse Functions

This is usually a troublesome topic for the student. One possible reason for some of the difficulty is simply the fact that this topic often is not introduced until near the end of the course—a time when interest often lags and time runs out. In any event, several authors have recommended introducing this important topic earlier and using the inverse function notation throughout.

Some debate has been staged over the respective merits of, for example, the notations of $\sin^{-1} x$ and $\arcsin x$. I would suggest using whatever notation the textbook uses, emphasizing at the same time that the other notation exists. Regardless of which notation is used, it is important to stress the reading of either one as "an angle (or number) whose sine is x." It is amusing and instructive to ask students for the value of $\sin (\sin^{-1} x)$. If they are still in the habit of reading (literally) sine of arcsine x or sine of inverse sine x, they are usually baffled. But if they read it as the sine of the angle whose sine is x, they get the point fairly quickly. (Although the teacher may have to lead them a little by asking, "What is the name of the man whose name is Smith?"!)

There is considerable difference of opinion (summarized in [R10]) as to how the principal values of the inverse trigonometric functions should be defined. It will do no harm, however, to use whatever definitions the textbook uses, providing that the students are cautioned that they may run across alternative definitions elsewhere. Common errors made by textbook writers on this topic are exposed in [L16]. For example, the author finds exercises such as, "Prove that arccos $(1 - 2m^2)$ $= 2 \arcsin m$." But the expressions on opposite sides of the identity symbol do not represent the same sequence of angles, while if we confine ourselves to the principal values of the angles, the identity does not hold for all values of m; for example, $m = -\frac{1}{2}$.

While many textbooks still treat the inverse trigonometric functions as multiple valued functions (e.g., $\sin^{-1}\frac{1}{2} = \pi/6$, $5\pi/6$ etc.) and call one of these values ($\pi/6$ in the example) the principal value, there is a growing trend to have, for example, $\sin^{-1}\frac{1}{2} = \pi/6$, $(\sin^{-1}\frac{1}{2}) + \pi = 7\pi/6$ etc. Thus we define $\sin^{-1} x$, for example, by $y = \sin^{-1} x$ if (1) $\sin y = x$ *and* (2) $-\pi/2 \leq y \leq \pi/2$. See, for example, [D19] or [S2].

While the use of the inverse function notation is extensive, we rarely find problems involving inverse functions like those stressed in some trigonometry textbooks (e.g., the calculation of $\sin (\cos^{-1}\frac{3}{5} + \tan^{-1}\frac{12}{5})$).

6. Trigonometric Equations

This topic provides an excellent review of most of trigonometry as well as a good deal of algebra but usually suffers from coming at the very end of the course. Generally speaking, student difficulties are those connected with background material rather than the specific topic of trigonometric equations.

Two methods of solving the important trigonometric equation $a \sin x + b \cos x = c$ are given in [K16] and in [W1]. The first applies only when a, b, and c are real and $a^2 + b^2 \geqq c^2$ and consists of introducing the half angle and dividing by $\sin^2 \frac{1}{2}x$ to obtain

$$\cot \tfrac{1}{2}x = \frac{a \pm \sqrt{a^2 + b^2 - c^2}}{c - b}, \qquad c \neq b$$

or, if $c = b$,

$$\tan \tfrac{1}{2}x = \frac{a \pm \sqrt{a^2 + b^2 - c^2}}{b + c} = \frac{a \pm a}{2b}.$$

The second uses the substitution $t = \tan \frac{1}{2}x$ to obtain

$$\sin x = \frac{2t}{1 + t^2}, \qquad \cos x = \frac{1 - t^2}{1 + t^2}.$$

Another useful method for solving this type of equation is much favored by engineers. We let $k = \sqrt{a^2 + b^2}$, $\sin u = b/k$, $\cos u = a/k$, and have $\cos u \sin x + \sin u \cos x = \sin (u + x) = c/k$.

Trigonometric equations are infrequently found in advanced mathematics and applied work, except in analytic geometry where the intersection of polar curves is discussed. The equation just discussed seems to be the most important, but even it is used only occasionally. See [D19] for a fairly detailed treatment of trigonometric equations.

7. Miscellaneous

We mention first several articles on the exact values of trigonometric functions for special arguments. First, [D14] gives a simple construction to determine the values of the trigonometric functions of $\pi/8$ and $\pi/12$ directly, without recourse to the sum and difference or half-angle identities. Second, [O6] shows that if the degree measure of an angle x is a rational number, the only possible rational values of the trigonometric functions are $\sin x$ and $\cos x = 0$, $\pm\frac{1}{2}$, ±1; $\sec x$ and $\csc x = \pm1$, ±2; and $\tan x$ and $\cot x = 0$, ±1. Similarly, [C9] shows that if the radian measure of an angle x is a rational multiple of π, and $\tan x$ is rational, then the radian measure of x is an integral multiple of $\pi/4$. Finally, [H6] gives an elementary proof of the theorem that the value of $\cos(a°b'c'')$ where a, b, and c are integers is an algebraic number.

I think that students deserve to be shown how the tables of trigonometric functions and logarithms are calculated from series. It is even possible to give derivations for these series without the aid of calculus (under the usual assumption that the series exist). For a derivation by the use of De Moivre's theorem and the binomial theorem see, for example, [P1]. There is also an interesting derivation of infinite series for $\tan z$ and $\ln(1 - z)$ in [H21]. The author assumes $\tan z = \displaystyle\sum_{n=1}^{\infty} a_n z^n$ and uses $\tan x - \tan y = (1 + \tan x \tan y)\tan(x - y)$ to obtain

$$\sum_{n=1}^{\infty} a_n x^n - \sum_{n=1}^{\infty} a_n y^n \equiv \left(1 + \sum_{n=1}^{\infty} a_n x^n \sum_{n=1}^{\infty} a_n y^n\right)\left(\sum_{n=1}^{\infty} a_n(x - y)^n\right).$$

Then by first equating coefficients of powers of y and then of powers of x (plus the use of $\tan \pi/4 = 1$), we can obtain $a_1, \ldots, a_n$. The series for $\ln(1 - z)$ is obtained similarly by the use of

$$\ln(1 - x) + \ln(1 - y) = \ln(1 - x)(1 - y)$$
$$= \ln(1 - x - y + xy).$$

See also [W14].

8. Logarithms

These may be considered as mechanical aids to computation or from the theoretical point of view. The computational part demands only sufficient drill for mastery. Negative characteristics cause the main trouble, and the bar notation (e.g., as $\bar{1}.0137$) should be avoided in this connection.

Since, once the procedure in a problem has been set up, the actual calculations and use of the tables present little difficulty, it would help to have extensive drill on "set-ups," without having the students complete the problem. In this way they will more quickly come to a realization of, for example, the incorrectness of $\log (x^2 + y^2) = 2 \log x + 2 \log y$.

Some confusion has been observed in the definition of a mantissa (see [R8]). In [H5] we have a careful formulation of a definition as follows: If x is a positive number, then x can be written in only one way in the form $x = d \cdot 10^p$, where $1 \leqq d < 10$ and p is an integer, negative, zero, or positive. Then $\log_{10} x = p + \log_{10} d$. We call p the characteristic of $\log_{10} x$ and $\log_{10} d$ the mantissa of $\log_{10} x$.

On the question of the merits of cologarithms, [R9] has this to say on the basis of experiment:

1. The subject of cologarithms should not be required.
2. Where there is an abundance of time, cologarithms may be introduced as optional material.
3. No definite statement can be made regarding the relative merits of the use of cologarithms and the nonuse of cologarithms.

My own opinion is more definite: Cologarithms are not worth considering.

There has been considerable discussion in recent years on the best method to present the definition and theory of logarithms. It is generally agreed that, if the concept of an integral is available, the best way to proceed is to define $\log_e a = \int_1^a dx/x$. This is done in several calculus books such as [T2] and [T5]. Exponential functions can then be defined in terms of logarithmic.

It is also possible, as in the SMSG textbook [S16], to use the intuitive concept of area under the curve $y = 1/x$ from 1 to a and not assume any knowledge of the calculus. Still another method begins with the functional equation $f(xy) = f(x) + f(y)$ with certain end conditions. See, for example, [S19] and [S21].

The traditional approach, of course, is to base logarithms on exponents; the weakness lies in the fact that it is not easy to define a^x for x an arbitrary real number and prove the laws of exponents for arbitrary real number exponents.

Whatever approach is used it seems clear that great stress should be laid upon the fact that

$$a^{\log_a x} = x$$

as this is the rationale behind most logarithmic manipulations. Thus

$$2 \times 4 = 10^{\log_{10} 2} 10^{\log_{10} 4} = 10^{\log_{10} 2 + \log_{10} 4}.$$

See, for example, [D18].

Certainly, with the increasing use of slide rules and computing machines, the computational aspects of logarithms deserve less and less emphasis compared to the noncomputational aspects. For every student who will use logarithms extensively for computations, there are many more who will need to know something about the theory of logarithms. The really basic idea is that there is an isomorphism between the multiplicative group of positive real numbers and the additive group of all real numbers, so that multiplication of positive real numbers can be reduced to the addition of real numbers. Thus if $x \rightarrow \log_a x$ and $y \rightarrow \log_a y$, we have $xy \rightarrow \log_a x + \log_a y$.

We have already mentioned the calculation of logarithms by series in Section 5.7. A more elementary approach is as follows: To calculate, for example, $x = \log_{10} 3$ we first write $10^x = 3$ and observe immediately that $0 < x < 1$. From a table of square roots we obtain $10^{1/2} = 3.16$ so that we now know that $0 < x < 1/2$. Next, we observe that $10^{1/4} = (10^{1/2})^{1/2} = 1.78$ and hence conclude that $1/4 < x < 1/2$. Let us now tabulate what we have obtained, as well as the rest of the work.

$$10^x = 3 \qquad\qquad\qquad 0 < x < 1$$

$$10^{1/2} \cong 3.16 \qquad\qquad\qquad 0 < x < 1/2$$

$$10^{1/4} \cong 1.78 \qquad\qquad\qquad 1/4 < x < 1/2$$

$$10^{x-\frac{1}{4}} = \frac{3}{10^{\frac{1}{4}}} \cong \frac{3}{1.78} \cong 1.69 \qquad\qquad 0 < x - 1/4 < 1/4$$

$$10^{\frac{1}{8}} \cong \sqrt{1.78} \cong 1.33 \qquad\qquad 1/8 < x - 1/4 < 1/4$$

$$10^{x-\frac{3}{8}} = \frac{3}{(10^{\frac{1}{8}})^3} \cong 1.28 \qquad\qquad 0 < x - 3/8 < 1/8$$

$$10^{\frac{1}{16}} \cong \sqrt{1.33} \cong 1.15 \qquad\qquad 1/16 < x - 3/8 < 1/8$$

If we stop our iterative process at this point we have $7/16 < x < 1/2$.   Averaging $7/16$ and $1/2$, we obtain $x \cong 0.46$.

Additional references on the theory of logarithms are [D15], [F2], and [H26].

6

The Teaching

of Analytic Geometry

Of recent years there has been a considerable trend toward the teaching of analytic geometry in combination with the calculus. In 1952, when the first draft of this handbook was written, I was all in favor of this trend. Experience, however, has led me to question its desirability, and I am not alone in this feeling. See, for example, [L10], and note that a report [C18] of the Panel on Physical Sciences and Engineering of the Committee on the Undergraduate Program in Mathematics has recommended that " . . . all calculus prerequisites, *including analytic geometry*, be taught in high school . . . " [my italics]. In any event, analytic geometry will be discussed here from a non-calculus viewpoint for convenience. We will not, however, discuss the subject of tangents, which certainly is a part of calculus whether we use the terminology of calculus or not. Also it seems a waste of time to treat any other than the simpler curves in a straight analytic geometry course, since the job of graphing is frequently made easier by the application of the calculus.

In the line of general remarks we first mention the unusual

book, [M26], which employs a vector treatment throughout. I believe, however, that this method of approach is too difficult for the average class. (See, however, a defense of the vector approach in [B31].) Nevertheless, I do strongly recommend that the vector point of view be introduced and that the student become familiar with the use of vectors. Certainly vectors are of very considerable use in many applications of mathematics.

It has also been suggested that proofs independent of quadrant can be made by an earlier introduction of translation and rotation. See, for example, [K8] and [M8]. To connect solid analytic geometry with plane analytic, the introduction of direction cosines in the plane has been advocated in, for example, [C21] and [W24].

It should also be mentioned at this point that many textbooks do not give enough attention to exceptional cases. Thus they state formulas such as $m = \dfrac{y_2 - y_1}{x_2 - x_1}$ and $\tan \theta = \dfrac{m_2 - m_1}{1 + m_1 m_2}$ without stating under what conditions these formulas hold. Subject to a similar criticism are textbooks which state that the slope of vertical lines is infinite rather than undefined (cf. remarks in Section 5.1 on $\tan 90°$, etc.).

The uses of specific parts of analytic geometry are summarized in the respective sections. Here we call attention to [L13], which deals with applications of analytic geometry to aircraft design and drawings. Analytic geometry might also be a good place for prospective users of mathematics in the sciences to study the relationship of mathematics to the physical world. See, for example, the articles [B15], [B26], [B28], and [H18], previously mentioned in Section 2.4.

1. Preliminaries

The first part of analytic geometry is usually devoted to the introduction of rectangular coordinates, distance between two points, division of a line segment, inclination and slope, and the angle that one line makes with another.

Only the division of a line segment causes any general difficulty. Here there is always some confusion regarding the difference between, say, dividing a line segment in the ratio 2 to 3 and in locating a point 2/3 of the way from one end to the other. Points of external division are also troublesome, and it seems doubtful if they are worth more than a mention, since application of points of division usually involve only internal division of a line segment.

Distance and slope, of course, are very frequently used. The angle between two lines (other than 0° or 90°) is used to a limited extent only.

2. The Straight Line

The only difficult parts of the usual chapter on the straight line are those connected with the normal form or with those problems for which no standard form is available (e.g., when the normal intercept and a point on the line are given). The latter type of problem is probably best treated from the viewpoint of families of lines. Thus, for example, if we have given that a line has slope 2 and that the product of its intercepts is $ab = 10$ we first write an equation, $y = 2x + b$, of the family of lines with slope 2. Then we seek the member or members of this family with $ab = 10$. Or we begin with the family

$$\frac{x}{a} + \frac{y}{10/a} = 1$$

and seek the member or members of this family with slope 2.

We may also note that problems involving the normal form are extremely rare outside of a course in analytic geometry. For this reason, as well as because of the difficulty students have with this topic, several writers have advocated the entire omission of the normal form, and some textbooks, at least, do so. On the other hand, the normal form is easy to use in the vector approach.

If the normal form is omitted, we must, of course, use a different derivation than that commonly given for the distance from a point to a line. Derivations not involving the normal form have been given in [M6], [B41], and [B3]. In the first derivation we consider the given line, l, with equation $ax + by +$

$c = 0$ and the point $P:(x_1,y_1)$.   Then the line through P parallel to l has the equation $ax + by + k = 0$ where $k = -ax_1 - by_1$. (See figure below.)   Now $d = PM = RN$ and $\dfrac{RN}{RR'} = \dfrac{OS}{SR}$,

$d = \dfrac{(RR')(OS)}{SR}.$ Then

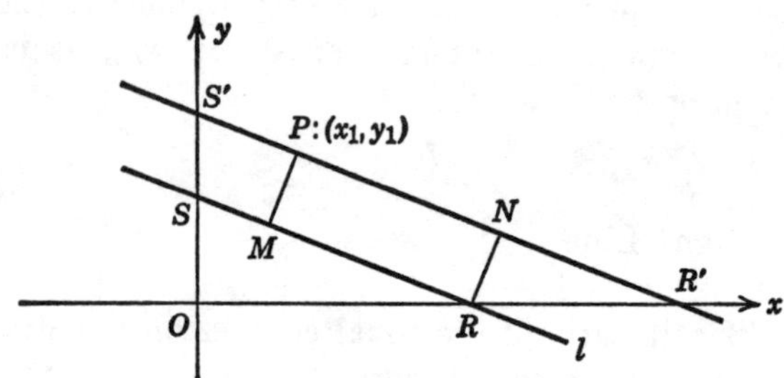

$$RR' = OR' - OR = -\frac{k}{a} + \frac{c}{a} = \frac{c - k}{a},$$

$$OS = -\frac{c}{b}, \quad SR = \pm \frac{c}{ab} \sqrt{a^2 + b^2},$$

$$d = \left(\frac{c - k}{a}\right)\left(-\frac{c}{b}\right)\left(\pm \frac{ab}{c \sqrt{a^2 + b^2}}\right) = \frac{ax_1 + by_1 + c}{\mp \sqrt{a^2 + b^2}}.$$

In the second derivation we take the given line with equation $y = mx + b$.   Then the line through P parallel to the given line has equation $y = mx + b'$.   From a figure, $d = (b' - b) \cos \alpha = \dfrac{b' - b}{\sqrt{1 + m^2}},$ where $\alpha = \tan^{-1} m$.   The third derivation is based on the following figure:

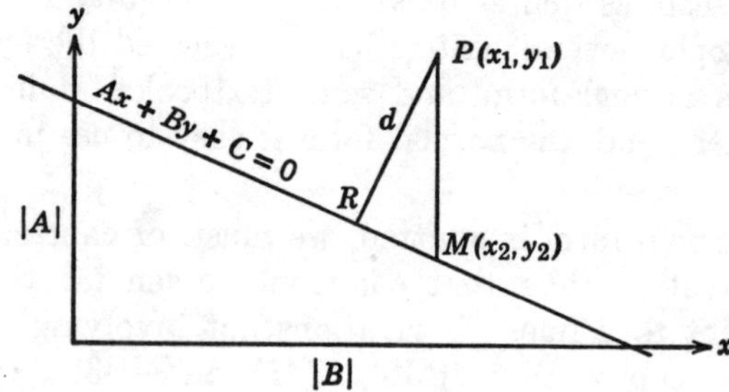

We have, since $x_2 = x_1$,

$$MP = y_1 - y_2 = y_1 - (-Ax_1 - C)/B = (Ax_1 + By_1 + C)/B,$$
$$RP/MP = |B|/\sqrt{A^2 + B^2}$$

from which

$$d = \frac{|B|(Ax_1 + By_1 + C)}{B\sqrt{A^2 + B^2}}.$$

Students and often teachers do not realize the complexities involved in the proper choice of sign for distance in analytic geometry. This matter is thoroughly discussed in [B21]. Three types of treatment are described:

(1) that where only the positive sign is used;
(2) that where both positive and negative signs occur, but the choice in any given instance is determined;
(3) that in which the sign is indeterminate, and both possibilities as to sign are retained throughout.

The author argues for the rejection of (2) as neither simple nor tenable in favor of either the obvious but difficult (1) or the simple but perhaps overinclusive (3).

While the formula for distance from a point to a line is occasionally encountered in advanced mathematics and in applications, the difficult application of this formula to the bisector of an angle is of interest only as an exercise in analytic geometry. This topic should, I believe, be treated lightly, if at all.

In passing, note that the formula for the distance from a point to a line is universally ascribed to Lacroix, who first published it in 1797. However, [B34] shows that Clairaut, in 1731, essentially gave this formula.

3. The Circle

Difficulties arise in the chapter on circles only when "miscellaneous" problems such as "find the equation of the circle passing through (2,1), tangent to the line $2x - y = 4$, and with center on the line $x + y = 5$" are considered. Again, as for similar problems on the straight line, such problems are best handled from the point of view of families of curves.

4. Conics—Special Cases

Under this heading we consider the usual treatment of the parabola, ellipse, and hyperbola as special curves rather than representatives of a class of curves known as conics. Student difficulty is confined largely to keeping in mind the terminology and the formulas needed. Students may be encouraged to master this material if they are told that this vocabulary is used quite extensively in such applied fields as atomic theory, optics, and mechanics. Usually in such applications the conics are found in standard position or, at worst, translated from standard position.

Most textbooks give a number of the usual applications of the conics. An unusual one is given in [P16]. The author shows that the form of the curve of intersection of an eroded slope and the alluvial fan of the plain is a conic. Not exactly an application, but an interesting problem in the theory of the ellipse, was proposed in [K12]. The author asked "How can one convince a class in analytic geometry that if the inside of a track is a noncircular ellipse, and the track is of constant width, then the outside is not an ellipse?" A solution is given in [B4]. For applications see also the book, [L13], mentioned in the beginning of this chapter. In regard to the names "ellipse," "parabola," and "hyperbola," see [E9].

An extensive literature exists on the subject of conic construction and graphing. We give a few references for these topics here.

1. There is a ruler and compass method for the point-by-point construction of central conics by ruler and compass in [H7].
2. [B7] utilizes the optical property of a parabola for a construction. That is, with O as focus we draw any line OC inter-

secting the directrix at C and from the point of intersection
construct CD perpendicular to the directrix. Then we con-

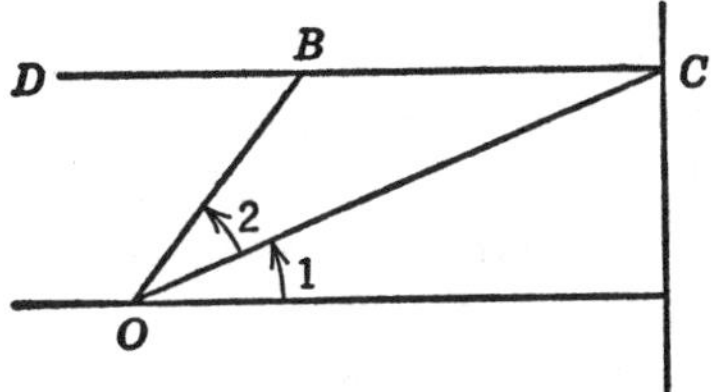

struct $\angle 2 = \angle 1$. The point of intersection, B, lies on the
parabola.

3. [V3] describes an ellipsograph as shown below. Here AC and
BC' are two unequal arms rotating in opposite directions from
the positions RC and SC' with equal angular velocity, and a

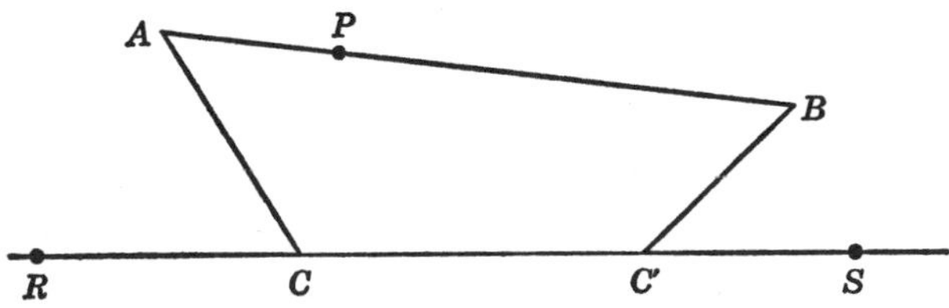

point P is on AB or AB extended so that $AP/PB = \lambda$ (a constant). The locus of P is an ellipse.

4. [B2] proves the following interesting theorems: Given a triangle, its base fixed in magnitude and position. Then the
loci of the points of intersection of the (a) medians, (b) per-
pendiculars constructed at the midpoints of the sides, (c) bi-
sectors of the angles, and (d) altitudes of all triangles having
the same area on the same base are, respectively, (a) a
straight line, (b) a straight line, (c) an ellipse, (d) an hyperbola.

5. [Y2] gives methods and proofs of the validity of the methods
for constructing the conics by paper folding.

6. [S43] describes an instrument for ellipse and hyperbola construction.

7. [J7] gives a construction for locating the point at which any
arbitrary line through the focus intersects the desired conic.

5. Translation and Rotation

We have already commented that these operations could be used more extensively if introduced earlier. Translation, of course, is simple, but rotation needs extensive drill before mastery. These operations are rather infrequently used outside a course in analytic geometry. Throughout the treatment of these topics, the idea of invariants should be stressed.

[S41] gives a rotation formula which does not involve trigonometry. Thus if $y - mx = 0$ and $x + my = 0$ $(m > 0)$ are a new pair of axes we have

$$x' = \frac{x + my}{\sqrt{1 + m^2}}, \, y' = \frac{y - mx}{\sqrt{1 + m^2}}.$$

6. Conics—General Case

This topic is often omitted or dealt with very sketchily in most of the analytic geometry courses of today. There is no doubt that considerable time is required for a meaningful explanation, and there is certainly no lack of different approaches to this problem. Some references are as follows:

1. [W20] tackles the problem directly, for nothing is assumed except the general definition of the conics.
2. [B42] gives a method of presenting the general conic theory, with considerable emphasis on invariants.
3. [M4] gives the details of a fairly standard presentation.
4. [Y1] uses the definitions: (a) a parabola is such that one, and only one, line of a pencil cuts the curve in only one point; (b) the hyperbola is such that there are two, and only two, distinct lines; (c) the ellipse is such that there is no line, to obtain the usual rules of classification.
5. [J5] gives a concise and elegant discussion with examples. This discussion is extended in [B24].
6. [F2], pp. 114–116 has a derivation by the method of Appolonius and, on pp. 117–118, by the method of Dandelin.

7. [H30] shows (geometrically) the equivalence of the two definitions of the conic sections as plane sections of a right circular cone and the corresponding geometrical loci (see also [K7]). This equivalence is also demonstrated analytically in [L5].

An additional reference is [H11].

7. Graphing and Loci Problems

Much of the graphing demanded of the student in analytic geometry books, when done without the aid of calculus, comes under the heading of cruel and unusual punishment, except for those students who like graphing.

In my opinion, from the point of view of applications, it suffices (in rectangular coordinates) to confine attention to graphs of polynomial functions of degree ≤ 4 plus the graphs of trigonometric, logarithmic, power ($y = x^n$, n an integer), and exponential functions, the conics, and the cycloids. A thorough study of these important curves would seem to me to be of much more value than the mechanical construction of a score of more complicated curves. It has also been suggested in [C6] that, for work in engineering, students should have some work with log log and semilog graph paper. Further work on more complicated curves can profitably be delayed, in my opinion, until the student has the use of the differential calculus.

Students who ask for the meaning of isolated points and complex, nonreal points of intersection may profitably be referred to [L2]. Here three dimensions are used with one axis for the real values of the independent variable and the other two axes for the real and imaginary parts of the dependent variable.

Most textbooks on analytic geometry confine work on polar coordinates to one chapter, but a few consider the polar forms of the various curves along with the Cartesian forms. Whatever scheme is used, we should be careful not to give the impression that the polar coordinate system is merely an afterthought.

Generally, the main difficulties in the plotting of polar curves arise in the trigonometry involved. It is a good idea to preface the work with polar coordinates with a review of the finding of functional values for angles of degree measure greater than 90 (see Section 5.3). Polar coordinates find extensive use in calculus. Outside calculus their use is confined largely to the polar form for the conics.

Loci problems are often difficult for students, but I have found no suggestions on the presentation of this topic. Several writers, however, have pointed out the common neglect on the part of textbook writers in proving the converse of a locus problem. That is, all too often they simply prove that if a point is on a certain locus, its coordinates must satisfy a certain equation and do not show that, conversely, if the coordinates of a point satisfy the equation, it lies on the locus. There seem to be few applications of loci problems in the sciences except for straight lines and the conics.

8. Solid Analytic Geometry

A small section of solid analytic geometry is customarily included in courses in analytic geometry, and it certainly should be. There is no need, however, to present a formal treatment. For applications to calculus and elsewhere it is sufficient for the student to have a working knowledge of direction cosines and numbers and that he be able to recognize equations of planes, surfaces of revolution, cylinders, and quadric surfaces as well as be able to sketch these figures. Less frequently one finds a use for the formula for the distance from a point to a plane and the angle between two lines in space.

An additional reference for this chapter is [Y4].

7

The Teaching

of Differential

Calculus

A sharp division of calculus into differential and integral is not implied here, and, in fact, I am much in favor of the presentation used in textbooks that alternate work in differential calculus with work in integral calculus. Our division here is simply to make it easier to locate material. The general remarks given here apply to both Chapters 7 and 8.

The controversy in regard to rigor, mentioned in Section 3.3, rages especially fiercely with respect to the calculus. There is general agreement, however, that the principles rather than the manipulational aspects of calculus are the important features not only for the prospective mathematician, but also for the prospective engineer, physicist, etc.

The ultimate aim, then, is the same for all, but the question is how to achieve this aim. Articles advocating approaches rather diametrically opposed to each other are [B8] and [M27].

The former article was mentioned in Section 3.3.   Briefly, it calls
for a psychological approach rather than a strictly logical approach.   [M27], on the other hand, demands that calculus
courses be started with a definition of real numbers as nested
sequences of intervals and continue at somewhat that level of
rigor.   Standing somewhat at the middle point in this controversy is [M1], which urges a compromise between the "problem
school" and the "ϵ and δ school."   See also [M25] in which,
among other things, it is urged that more differential geometry
be injected into beginning calculus.

There are textbooks available today to fit just about anyone's
taste for rigor or lack of rigor.   Actually, the determining factor in the selection of texts should be the abilities of the students, rather than the viewpoint of the instructor.   The instructor (especially if he is new to teaching) will usually be biased in
favor of the more rigorous presentation.   If you find yourself
inheriting an inadequate textbook, be careful not to discard it
entirely in favor of a lecture presentation.   As I mentioned before, the student at this level is not too well prepared to take
adequate lecture notes in mathematics.

It is certainly true that calculus is a much applied subject.
However, it has been repeatedly stressed that successful application rests on an understanding of fundamental principles.   Every
effort should be made to promote an intuitive feeling for the fundamental concepts of the calculus, so that the engineer and
others will turn naturally to calculus when faced with a new
problem.   Only too often scientists regard it as a strange, exotic
thing that ordinary persons like themselves cannot use.   An
overemphasis on "pathological" aspects of calculus for this type
of student will be a real deterrent.   See, for example, [P14].

1. Limits

It has been argued in [B8] and elsewhere that the first day or so
of teaching calculus can most profitably be devoted to an explanation of the reasons why we are "forced" to study calculus
through consideration of the velocity at a point, or the tangent

to a curve. Sooner or later, however, the concept of a formal limit must be faced. To my mind it is best to treat this difficult topic from an intuitive point of view in the beginning, and to return to the topic from time to time with a more mature treatment. Thus the primary goal in beginning calculus should be to give the student a feeling for limits rather than familiarity with "epsilonics," and this goal is probably best met by the consideration of a host of examples (see [K6]).

The important special limit, $\lim\limits_{\theta \to 0} \dfrac{\sin \theta}{\theta} = 1$, is considered in [K19] as follows: If P is the perimeter of a regular n-gon inscribed in a circle of radius r, then $P = 2nr \sin (\pi/n)$, and we know from plane geometry that $\lim\limits_{n \to \infty} P = 2\pi r$. Hence $\lim\limits_{n \to \infty} (n/\pi) \sin \pi/n = 1$, and if we let $\pi/n = \theta$, then $\theta \to 0$ as $n \to \infty$ and conversely. Hence $\lim\limits_{\theta \to 0} \dfrac{\sin \theta}{\theta} = 1$. In [F10], on the other hand, the author defines $\sin x$ by a series and then gets the desired result via the general theorem "If $a_0 + a_1x + \cdots$ converges when $|x| < R$ to $f(x)$, then $\lim\limits_{x \to 0} f(x) = a_0$." For still another approach that avoids the concept of area see [H25].

2. Continuity

The subject of continuity forms another major stumbling block in the beginning of calculus. Again I feel that an intuitive approach based on the consideration of a large number of examples should precede a more formal approach. Thus I would recommend that the definition for continuity be phrased as follows:

The function f is said to be continuous at $x = a$ if and only if

1. $f(a)$ exists.
2. $\lim\limits_{x \to a} f(x)$ exists (from both sides).
3. $f(a) = \lim\limits_{x \to a} f(x).$*

*It should be mentioned, of course, that 1 and 2 are both implied by 3. But for pedagogical reasons it seems best to consider 1 and 2 separately and to consider 3 as simply questioning the equality.

Then numerous examples should be given where 1 fails but 2 holds, 2 fails but 1 holds, etc. In 2, of course, it must be pointed out by examples that the limit does not exist if it is not independent of the way in which x approaches a.

No consideration of continuity is complete, I think, without the mention of the classical discontinuous functions defined by

(1)
$$f(x) = 0 \text{ if } x \text{ is irrational,}$$
$$f(x) = 1 \text{ if } x \text{ is rational.}$$

and

(2)
$$f(x) = 0 \text{ if } x \text{ is irrational,}$$
$$f(x) = 1/q \text{ if } x = p/q \text{ in lowest terms.}$$

These, of course, should be recalled when integration is studied, since the function defined by (1) is about the "simplest" function that is not Riemann integrable, while the function defined by (2) is about as "bad" as a function can get and still be Riemann integrable.

Other approaches to continuity have been suggested. [J2] would say that "a function is said to be continuous in a given interval providing it is uniformly continuous over some set of subintervals which cover the entire interval." Similarly, [W19] defines a continuous function as one that possesses a limit at every point and in which the limit is approached uniformly.

Several writers have exhibited simple functions with finite discontinuities. Thus [W5] lets $y = $ degree in u of $(x - 1)u^3 + 5u^2 - 3u + 2$. Then $y = 2$ at $x = 1$ but otherwise $y = 3$. Other examples are $y = $ degree in u of $(x - 1)u^{x^2+4} + 5u^2 - 3u + 2$, $y = $ degree in u of $u^{x^2} + 2u^2 - 5u + 3$. In [C3] the author points out that, for example, $\displaystyle\lim_{x \to \sqrt{2}} a^x$ does not exist when $a < 0$, because if $x = 1.41, 1.414, \ldots$ we get some imaginary values for a^x while if $x = {}^{141}\!/_{99}, {}^{1414}\!/_{999}, \ldots$, we get real values for a^x, some positive and some negative. Also $\displaystyle\lim_{x \to 0} a^x = 1$ if $x \to 0$ through $\frac{1}{3}, \frac{1}{9}, \frac{1}{27}, \ldots$ while if $x \to 0$ through $\frac{1}{2}, \frac{1}{4}, \frac{1}{8}, \ldots$, a^x takes on imaginary values. Similarly, [G1] discusses the function $y = x^x$.

3. Definition of the Derivative

The desirability of beginning calculus with an intuitive discussion of a derivative from either the kinematic or the geometrical point of view has already been mentioned. Whether this is done or whether the consideration of the derivative is deferred until later in the course, it is certainly important that this basic topic be carefully introduced as a meaningful process rather than a mechanical procedure.

I usually begin by considering the problem of finding the slope at the point (1,1) of the curve $y = x^2$ by a heuristic argument. First of all, the students are invited to solve the problem on their own until they come to a realization of the inadequacy of the mathematical tools at their disposal.* Then we "give up" trying to find the tangent and seek the slopes of secants. Starting with the slope of the secant through (1,1) and (2,4) we work down to, say, the point ($\frac{5}{4}$, $\frac{25}{16}$). Similarly we start with the secant through (1,1) and (0,0) and work up to, say, the point ($\frac{3}{4}$, $\frac{9}{16}$). Then, reviewing the concept of a limit, the students quickly agree that the slope of the tangent must be 2.

This is followed by a more rigorous argument employing $1 + h$, and, finally, the same argument is generalized to the arbitrary point (x_0, y_0) on $y = x^2$. A similar discussion, of course, can be made using the velocity of a freely falling body. One of these detailed discussions should be presented, I think, before work proceeds on formal definitions and derivations of formulas.

4. Differentiation Theory

The general theory of differentiation concerns itself with the rules that are independent of the functions involved. These are the sum, product, and quotient rules and the composite function rule.

---

*I have never yet had a student suggest using $(y - 1) = m(x - 1)$ and finding the line of the family making single contact with the curve $y = x^2$. If such a suggestion is made, it can easily be rejected as being insufficiently general for our purposes.

The special theory of differentiation concerns itself with the development of the rules for differentiating particular functions. No discussion of algebraic functions is needed here, but several references can be given concerning transcendental functions. Thus [H23] first finds the derivative of $\tan x$ by writing $\tan (x + \Delta x) - \tan x$ as

$$\frac{\sin (x + \Delta x)}{\cos (x + \Delta x)} - \frac{\sin x}{\cos x} = \frac{\sin \Delta x}{\cos (x + \Delta x) \cos x}$$

or as

$$\frac{\tan x + \tan \Delta x}{1 - \tan x \tan \Delta x} - \tan x = \frac{(1 + \tan^2 x) \tan \Delta x}{1 - \tan x \tan \Delta x}$$

$\left(\text{and using } \underset{\alpha \to 0}{L} \frac{\tan \alpha}{\alpha} = 1\right).$ Then $\tan^2 x + 1 = \sec^2 x$ gives the derivative of $\sec x$; $1/\sec x = \cos x$ gives the derivative of $\cos x$, etc. There is also a modification of the usual procedure of first finding the derivative of $\sin x$ given in [F10], in which $\sin (x + \Delta x) - \sin x = 2 \cos (x + \Delta x/2) \sin (\Delta x/2)$ is used.

Since differentiation of the logarithmic and exponential functions is based on the existence of $\underset{n \to \infty}{L} (1 + 1/n)^n$, our references and comments will deal with this point. A heuristic approach to e is given in [R4]. The author begins by finding the slope of $y = 2^x$ as 0.70 at $(0,1)$ by interpolation, using a four-place table. Similarly, the slope of $y = 3^x$ is calculated as 1.10. This suggests asking whether or not there is some number for which the corresponding slope is 1. Assume there is, and call it e. Thus when Δx is the smallest value for which we can interpolate, we shall have $y = e^x$ yielding $\Delta y/\Delta x = 1$. For Δy we have $e^{\Delta x} - 1$ at $(0,1)$ and therefore must solve $(e^{\Delta x} - 1) \Delta x = 1$ or $\Delta x \log e = \log (1 + \Delta x)$. From a seven-place table we have $\log (1 - 0.001) = -0.0004345$, $\log (1 + 0.0001) = +0.0004341$, which suggests that when Δx is nearer to zero than ± 0.0001, we should use 0.0004343. Then $\log (1 + \Delta x) = 0.0004343(\Delta x/0.001)$, and therefore $\log e = 0.4343$, $e = 2.718+$.

Formal proofs of the existence of the required limit are available in many texts. See also, [D23], [H34], [L17], and [P13]. Also

see the derivation of differentiation formulas for the trigonometric, exponential, and logarithmic functions from their
series expansions in [E2], and note that if ln a is *defined* as
$\int_1^a dx/x$, $(d \ln a)/da = 1/a$ (see Section 5.8) follows immediately
from the fundamental theorem of the calculus.

Under the heading of miscellanea, we note the rather amazing
simple approximate construction for e given in [S1]. The author
takes an equilateral triangle ABC with the altitude AD bisected
at E. The line BE is produced to F and then to G so that
$BF = 2BC$ and $FG = \frac{1}{8}BE$. The points H and I are taken on
BE so that $BH = BD$ and $BI = \frac{1}{3}BC$. The point J is taken on
BC so that HD, IC, and JG are concurrent. The side BC is
produced to K so that $CK = 2BC$. Then

$$\frac{JK}{BC} = \frac{228 + 11\sqrt{3}}{84 + 4\sqrt{3}} = 2.718,281,828(30)$$

with an error of approximately 0.000,000,000,16. Also under
miscellanea we have the delightful quote from the well-known
naturalist, J. Henri Fabre, on the appearance of e in nature
(given in [D7], pp. 180–181) and the Hermite proof that e is
irrational in [F2].

5. Differentiation Mechanics

With sufficient drill even the poorer students can become quite
proficient in differentiation. The teacher must be prepared, however, to spend considerable time aiding the weaker students to
review the algebra and trigonometry involved. Special attention
needs also to be paid to the rule for differentiating a composite
function. I find that it helps to use the suggestive words "interior" and "exterior." For example, in

$$y = \sin (x^2 - 2)$$

I speak of $x^2 - 2$ as the interior and sin as the exterior. Thus
y' is the derivative of the exterior times the derivative of the

interior,

$$y' = [\cos{(x^2 - 2)}](2x) = 2x \cos{(x^2 - 2)}.$$

I also find it desirable to emphasize the fact that any "elementary" function can be differentiated by repeated application of simple rules by going through the differentiation of a few pathological monstrosities such as

$$y = \sin{[\log{(\cos{e^{x^2-1}})}]}$$

where sin is the first exterior and $\log{(\cos{e^{x^2-1}})}$ the first interior; log is the second exterior and $\cos{e^{x^2-1}}$ the second interior, etc. At the same time it should be noted that most applications of differentiation are almost entirely confined to relatively simple functions, so that extensive drill in the differentiation of complicated functions is not justified.

6. Second and Higher Derivatives

Under this heading I mention three articles. The first, [R2], gives a geometrical meaning for $f''(x)$ by showing that $f''(x)$ is the curvature at its vertex of the vertically symmetric parabola which most closely fits the curve $y = f(x)$ at the point whose abscissa is x.

The second, [F1], observes that the general principle involved in the simplification of a second derivative of a function by the use of the original function is as follows: If $F(x, y)$ is a homogeneous function of order n in x and y, then $F(x, y)$ is a factor of the second derivative of y with respect to x of the equation $F(x, y) = c$ (Euler's theorem on homogeneous functions).

The third article, [I2], takes $f = g/h$ and derives the formula

$$f^{(n)} = \frac{1}{h^{n+1}} \begin{vmatrix} h & 0 & 0 & \cdot & 0 & g \\ h' & h & 0 & \cdot & 0 & g' \\ h'' & 2h' & h & \cdot & 0 & g'' \\ \cdot & \cdot & \cdot & \cdot & \cdot & \cdot \\ h^{(n)} & \binom{n}{1}h^{(n-1)} & \binom{n}{2}h^{(n-2)} & \cdot & \binom{n}{1}h' & g^{(n)} \end{vmatrix}$$

by solving for $f^{(n)}$ in the system of equations

$$hf = g$$

$$h'f + hf' = g'$$

$$\cdots\cdots\cdots$$

$$h^{(n)}f + \binom{n}{1}h^{(n-1)}f' + \binom{n}{2}h^{(n-2)}f'' + \cdots + hf^{(n)} = g^{(n)}.$$

This result, of course, is not of great practical importance. It is rather rare that derivatives of higher order than the second are used, and even those of the second order are confined largely to simple functions.

7. Maxima and Minima of Functions of One Variable

When these problems are encountered in connection with graphing, they cause little trouble. As usual, difficulties arise in the translation of the word problems into symbolic notation, and this is frequently a problem in algebra or trigonometry rather than in calculus. Hence it is suggested that the material on word problems given in Chapter 4 be reviewed at this time.

Several authors have complained about the lack of rigor in the usual presentation of maxima and minima. A treatment of the logic underlying this topic is given in [W3]. In this article the author analyzes the simple problem: "A piece of tin three feet square is to be made into a rectangular box open at the top by cutting out a square of tin from each corner and bending up the sides of the resulting pieces parallel with the edges. Among all such boxes, to find the box of greatest volume." The logic underlying the formal work is analyzed as follows:

1. The function V defined by $V(x) = x(3 - 2x)^2$ is continuous in the closed interval $0 \leq x \leq \frac{3}{2}$, hence possesses a maximum there. (It is assumed that any function continuous in a closed finite interval possesses a maximum.)
2. The function V vanishes for $x = 0$ and $x = \frac{3}{2}$ and is positive

between these values, so the maximum occurs at some point or points in the *interior*, $0 < x < \frac{3}{2}$, of the original interval.

3. At the maximum of $V(x)$ we must have $D_x V = 0$. (If $D_x V > 0$, the function V increases as x increases; if $D_x V < 0$, the function V decreases as x increases and increases as x decreases; in either case V has no maximum.)

4. Among the values $0 < x < \frac{3}{2}$ we have $D_x V = 0$ if and only if $x = \frac{1}{2}$.

5. Hence $x = \frac{1}{2}$ yields the maximum value of V.

See also [R19], [M2], [O1], and [F10], pp. 124–127, which discuss functions that have a maximum or minimum at end-points but no derivative equal to zero. For example, in [O1] we have the problem: "Given a straight fence 100 feet long, to add 200 feet more to form a rectangular enclosure whose boundary contains the original fence and will enclose the greatest area." The formal approach yields $A = -x^2 - 50x + 5{,}000$ where x denotes the length of new fence which is aligned with the original 100 feet of fence. Then $D_x A = -2(x + 25) = 0$, $x = -25$. The correct solution is, of course, $x = 0$, which is an endpoint minimum. Similarly, [R19] discusses the problem: "A man is in a boat 5 mi. from a straight shore line. He can row at the rate of 3 mi. per hour and walk at the rate of 4 mi. per hour. How far from the point on the shore directly opposite him should he land to reach a point s miles down the shore in the shortest time." Here, if $s < 15/\sqrt{7}$ the usual procedure does not give the minimum. [M2] gives the theory underlying such problems.

Most proofs of the existence of a maximum value involve the concept of a least upper bound, but in [F8] there is a proof of the theorem that if a function is continuous on a closed interval I, there exists a maximum on I which does not involve this concept.

[B17] points out that not all maxima-minima problems can be handled by elementary calculus (as, for example, the brachistachrone problem) and suggests that this point be made clear to the class.

Applications of maxima and minima appear moderately frequently and are usually of a relatively simple nature. Hence it would appear that some of the more involved problems customarily given are of little practical value.

8. Graphing

This topic has already been discussed in Section 5.7 where it was pointed out that it seems foolish to spend excessive time on graphing without the aid of the calculus. Even with the aid of the calculus, however, many of the problems customarily considered are very time consuming. Again I would say that it would seem a better procedure to analyze a few simple curves thoroughly than to give superficial consideration to a lot of more complicated curves.

It has also been suggested that graphing be motivated by asking for numerical results. That is, for example, instead of simply asking for the graph of $y = x^3 - 2x^2 + x - 1$, ask for the zeros of $x^3 - 2x^2 + x - 1$ or its points of intersection with the line $y = x$. This is actually how much graphing is used in practice.

Again, rigorous proofs of the theorems relating to the use of the derivative in graphing are beyond the grasp of the average student. Such proofs are well presented in [T3].

9. Velocity and Acceleration

From the point of view of usefulness in physics and engineering, this is one of the most important topics in the differential calculus. The calculus teacher, however, may best confine himself to emphasizing the basic ideas, since all books on mechanics have chapters on both rectilinear and curvilinear motion giving the details relating to applications.

[P2] derives formulas for the components of velocity and acceleration on the plane using complex number representation of vectors, and [Y3] obtains the normal and tangential acceleration directly in polar coordinates rather than by converting from a rectangular reference system.

10. Differentials

Very considerable objection has been raised to the customary presentation of differentials. However, there is little agreement as to what is the best presentation. Various points of view were expressed in 1942 in [K1], in [J1], and in [C10].

In the first article, the authors urge that we let each of the symbols $D_x y$, $D_x f$, $D_x f(x)$, y', f', and $f'(x)$ stand for

$$\operatorname*{L}_{\Delta x \to 0} \frac{f(x + \Delta x) - f(x)}{\Delta x},$$

but that, above all things, we do *not* define dy/dx to be this limit. Then we let dx be an absolutely arbitrary number and define

$$dy = f'(x)\ dx.$$

Looking at the geometric representation of dx, dy, Δx, and Δy, we note that *if* we take $dx = \Delta x$ and "small," then dy is an approximation to Δy.

In the second article, the author agrees with the suggestions in [K1] but adds the following summary to be presented "*after* the dust of construction and the smoke of discussion have been cleared away": If y is a function of x, in symbols, $y = f(x)$, dy *is a function of the two variables x and dx*, represented by the formula $dy = f'(x)\ dx$. Graphically, dy is the increment of the ordinate of the tangent line if x is given an increment Δx equal to dx. If dx is infinitesimal and $\Delta x = dx$ (and if each of the variables x, y has, with respect to the other, a derivative different from zero), dy differs from Δy by an infinitesimal of higher order than dx or dy or Δy.

In the third article, the author agrees with [K1] that the usual definition of the differential is unsound but does not agree that the objections that they make to it reveal the unsoundness of it. Furthermore, the author points out, the use of differentials in connection with integration is not provided for. Thus, if dx is a new independent variable, we certainly are at liberty to call it, say, z so that $\int x\, dx$ becomes $\int xz$, and the whole significance of the notion is lost. After some discussion along these lines the author states, "Unless some solution of these difficulties can be found, it seems that it would be preferable to introduce differentials in a frankly inaccurate and heuristic manner as small values of the increment of 'little bits' of the 'variable quantity' involved, rather than to clothe the idea with the deceptive appearance of logical accuracy." Finally, [C10] concludes that there is a method of introducing differentials which might remedy these difficulties, but it involves spending a disproportionate amount of time on the study of parametric equations. In this method we define $dx = D_t x$, $dy = D_t y$, where t is an arbitrary parameter.

In 1952 the whole subject was reopened in the *American Mathematical Monthly*. See [F9], [P5], and [H3]. In summarizing these, and earlier articles, the editor, C. B. Allendoerfer, comments that he " . . . is convinced that there is no commonly accepted definition of a differential which fits all uses to which this notation is applied." See also [M25] for a point of view from differential geometry, [R3] which advocates an earlier introduction of differentials, and [L15].

11. The Law of the Mean

I have previously commented (Section 3.3) that the usual proof of the law of the mean from Rolle's theorem by the unmotivated introduction of a function F that "works" has often been criticized. In [B19] the author points out that the ordinate difference between the curve and the chord (which becomes the ordinate under a simple transformation of axes) obviously satisfies the conditions required for F. Similarly, [S34] presents a motivating device by the use of Taylor's series, and [E8] and [W4] by use of translation-rotation of the coordinate system. In [P17], the author uses an analogous approach to [B19] to obtain the first mean-value theorem of the integral calculus.

The significance of the law of the mean is best understood by the beginner through applications. Some very interesting ones are given in [P13] and include the establishment of the existence of $\lim_{n \to \infty} \left(1 + \dfrac{1}{n}\right)^n$, the equalities $\sin x = x - \epsilon$ $(\epsilon < x^3)$,

$\cos x - 1 = -x^2/2 + \epsilon \, (\epsilon < x^4/4, x < 1), (1 + x)^{1/n} = 1 + x/n$ $- \epsilon \, (\epsilon < x^2/n),$ $\tan^{-1}x = x - \epsilon \, (\epsilon < x^3)$ and the justification of the ordinary rules of interpolation in tables of sines and cosines (the latter requiring a double application). In connection with the last named application see also [H33]. This shows that if f is a function which, together with its first derivative, is continuous throughout the interval $a \leqq x \leqq b$, and $f''(x), f'''(x)$ are of constant sign throughout the interval, then the error of interpolation for $f(x)$ from $f(b)$ to $f(a)$ is less than or equal to $\frac{1}{2}$ $(b - x)(x - a)f''(\theta)$ and $|\text{error}| \leqq \dfrac{(b - a)^2}{8} |f''(\theta)|$ where $a < \theta < b$.

12. Indeterminate Forms

First of all we mention that in [F10] the author objects to the use of the terminology "indeterminate" forms and prefers to call them simply limits. Students generally find this topic interesting, and it affords an excellent opportunity for a review of limits. It does not seem to be applied to any great extent, and the wisdom of spending much time on the more complicated indeterminate forms is questionable.

The main difficulty to guard against is the student's tendency to apply the rule of differentiating both numerator and denominator mechanically, without stopping to verify at each step that he has either a 0/0 or an ∞/∞ situation. In [O8], by the way, it is pointed out that almost all problems in existing textbooks of the form ∞^0 have a limit equal to 1, and examples are given where the limit is not 1 (e.g., $\mathop{L}\limits_{x \to \infty} (e^x + x)^{1/x} = e$).

The list of examples should certainly include one (as $\mathop{L}\limits_{x \to \infty} \left(\dfrac{x - \sin x}{x} \right)$ for which a limit exists, but for which L'Hospital's rule provides no evaluation. Students may also be interested in some geometric representations of indeterminate forms given in [M20].

[T4] gives a proof of L'Hospital's rule which " . . . has the virtue of disposing completely of the rule in brief compass, without any need for tedious change of variables or the treatment of a variety of special instances." It is based on Cauchy's formula (the extended law of the mean). Other proofs are also given.

Along with a discussion of indeterminate forms should go some work on orders of magnitude at both 0 and ∞. Thus, for example, we have both

$$\mathop{L}_{x \to \infty} x = \infty \qquad \text{and} \qquad \mathop{L}_{x \to \infty} e^x = \infty,$$

but

$$\mathop{L}_{x \to \infty} \frac{x}{e^x} = 0.$$

Thus x is of a lower order of infinity than e^x, but, since

$$\mathop{L}_{x \to \infty} \frac{\log x}{x} = 0,$$

x is of a higher order of infinity than $\log x$. Such considerations are frequently of considerable importance in applications.

Additional references for this chapter are [B36], [F7], [H36], [J6], and [R20].

8

The Teaching
of Integral
Calculus

Most textbooks begin integral calculus with a discussion of the anti-derivative (usually confined to polynomials), and some even go on to the finding of simple areas and volumes before bringing in the definite integral as a limit of a sum. To the student, the anti-derivative approach to the integral calculus is certainly the easier, but, on the other hand, the heart of the subject lies in the sum approach.

I see no reason why anti-derivatives should not be introduced first—probably in connection with the study of the motion of a particle. I do object, however, to using the anti-derivative for finding areas, etc., as such a procedure tends to make the student feel that the sum concept introduced later is unessential and unimportant. And while, eventually, the word integral will replace the word anti-derivative, I strongly recommend that the latter term be used exclusively until the sum concept is introduced.

Otherwise, to the average student, it appears as if the fundamental theorem of the calculus is simply stating that "an integral is an integral!"

1. The Definition of the Riemann Integral

The reader may recall that I suggested introducing the derivative (Section 8.3) by means of a heuristic argument from simple examples. I also recommend, such a procedure to introduce the integral beginning with, say, the finding of the first quadrant area under the line $y = x$ from $x = 0$ to $x = 1$ and concluding with the finding of the first quadrant area under the curve $y = x^2$ from $x = 0$ to $x = 1$. The value of the first example, of course, is that the result can be checked. In the second example, students should be made to meditate over the difficulties of the problem and made to see the *need* for a new tool. In both examples, upper and lower sums are calculated for equal divisions of the interval from 0 to 1 into, say, first 4 and then 10 parts, and an informal induction is made to the general case.

This should probably constitute the first day's work. Homework can consist of similar examples to be approached in the same intuitive fashion. Then, on the second day, the general definition can be given with illustrations drawn from the previous day's work. The exact level of the general presentation will, of course, vary according to the ability of the class. This concept of a definite integral is probably the most complicated notion that students have yet met with in their mathematics, and there is no point in giving a presentation pitched at so high a level that the entire class loses interest. Regardless of the level of presentation, I believe that it is vital to emphasize that the integral is defined without the use of the concept of area. Too often, students have the impression that the definite integral is defined as an area, rather than the other way around.

The usual presentation of the definite integral as a limit of a sum has been sharply criticized in [M27], since the student has studied only limits of functions of one variable, and the limit used here is certainly of a much more sophisticated type. In its place the author suggests introducing the usual upper (S) and lower (s) sums and then showing that S is bounded below with lower bound B, and s is bounded above with upper bound b. If $B = b$, the function is said to be integrable over the interval in question, and the common value of B and b is said to be its integral over the interval. Similarly, [A1] criticizes the partitioning procedure and defines the integral "rigorously and simply as an ordinary limit of an ordinary sequence of numbers." See also the approach given in [A8] via step functions.

The third day in the study of the definite integral should logically be devoted to a proof of the fact that the Riemann integral of a continuous function exists (for elementary proofs see, for example, [F10] or [S37]). For many (perhaps most) classes, however, such a presentation would be way beyond them and out of place. Whether or not the proof is given, the theorem should certainly be stated, and, in addition, certain other remarks definitely ought to be made. Thus it should be pointed out that it is sufficient but not necessary that a function be continuous in order for it to have a Riemann integral. As an extreme example, it might be mentioned (without proof) that the function f defined by $f(x) = 0$ if x is irrational, $f(x) = 1/q$ if $x = p/q$ in lowest terms is Riemann integrable. Along the same lines it should be shown that the function f defined by $f(x) = 0$ if x is irrational, $f(x) = 1$ if x is rational is not Riemann integrable. At this point, too, the existence of other integrals such as the Lebesque might well be mentioned.

So far I have mainly suggested additions to the usual textbook presentation. Perhaps a word of caution in the other direction is more important for the beginning instructor who is likely more than ready to discuss the complete theory of integration with the class. As in all cases where additional material is presented, the instructor should keep a close watch on class reaction. It is not necessary that the class understand all that is said, but if they show signs of definite lack of interest—it is a good time to stop.

The fourth day can be devoted to a review and an exposition of the theorems on the integral of a sum etc., and the fundamental theorem of calculus. Every effort should be made to make the students see the remarkableness of the latter. It will help to achieve this aim if, up to this point, the word anti-derivative (rather than integral) has been used for the inverse of differentiation.

See also [R23].

2. Approximate Integration

It seems to me that the subject of approximate integration naturally follows the introduction of the integral as a sum, and that increased stress should be placed on this topic in view of the wide use of computers. There is, of course, an extensive literature on numerical integration and we mention here only a few items clearly connected with the usual textbook treatment. First of all, a formula for the error in Simpson's rule is given in [R1], p. 384 as

$$E \leqq \frac{(b-a)^5 M}{180n^4},$$

where M is any constant such that $M \geqq |f^{(4)}(x)|$, $a \leqq x \leqq b$. The proof is given on pp. 461–464. More elaborate and precise formulas for this error are given in [S6].

In [D3], the author develops four rules like Simpson's rule from Euler's summation formula; and [R1], pp. 386–390 gives Gauss' rules. [K17] describes numerical integration in polar coordinates; [H9] provides a correction for the trapezoidal rule; and [M24] derives the Euler-Maclaurin formula for approximate integration from Taylor's formula.

3. Formal Integration

This topic is, I think, greatly overstressed in most courses. Generally speaking, the integrals that appear in applications are either fairly simple, or they can be handled only by approximate integration. The most important types of integrals seem to be those integrable by parts, those whose integrands are rational functions or powers of sine and cosine, and those handled by trigonometric substitution.

The literature on formal integration is rather extensive and we select only a few items for mention here. Under integration by parts, [R1] remarks that this is " . . . a market in which any integral may be traded in on another one; the trader must only be careful to obtain good integrals for poor ones!" [B30] notes that in using the formula for integration by parts, the constant of integration is invariably chosen as zero. That this is not always the best choice is illustrated by

$$\int x \tan^{-1} x \, dx = \tfrac{1}{2}x^2 \tan^{-1} x - \int \frac{x^2 \, dx}{2(1 + x^2)}$$

when we choose $c = 0$. But if we choose $c = \tfrac{1}{2}$, we have

$$\tfrac{1}{2}(x^2 + 1) \tan^{-1} x - \tfrac{1}{2}\int dx.$$

The importance of the constant of integration is vividly illustrated by a paradox given in [W2]. Thus, in the integration of $\int dx/x$ we let $u = 1/x$, $dv = dx$. Then $du = -dx/x^2$, $v = x$, $\int dx/x = 1 + \int dx/x$, $0 = 1$!

Under integration of rational functions we mention only [S35], which presents a device to simplify the work involved in the treatment of partial fractions with repeated linear or quadratic factors. It is worthwhile to stress the fact that if we assume the wrong decomposition of a given fraction into partial fractions, there will usually be nothing to show us our error unless we check. This provides an excellent illustration of the necessity for an existence proof.

The integrals of trigonometric functions can get quite difficult. For most practical purposes, however, it suffices to consider $\int \sin^n x \, dx$ and $\int \cos^n x \, dx$ for n even and positive and $\int \sin^n x \cos^m x \, dx$, where m or n is an odd positive integer.

The use of trigonometric substitutions to eliminate algebraic irrationalities is largely confined in practice to simple varieties which are commonly handled by tables. And this leads to a final remark on formal integration. Certainly at least one class session should be devoted to the use of tables of integrals, and students should be informed of the existence of more extensive tables than those in their textbooks (e.g., [D24] and [P4]). Along with this should certainly go some examples of functions that are not integrable in terms of elementary functions, together with a brief discussion of what it means for a function to be integrable in terms of elementary functions.

4. Improper Integrals

The type of improper integral with infinite limits occurs quite frequently in applications—in electricity, optics, mechanics, physical chemistry, and, of course, kinetic theory. On the other hand, the type of improper integral which has a discontinuity of the integrand within the limits of integration occurs much less frequently.

Because of its practical importance it might be well to evaluate $\int_{-\infty}^{\infty} e^{-x^2}\, dx$. A simple treatment is given in [S2] as follows. If we let $\theta = \int_0^{\infty} e^{-x^2}\, dx$, we have

$$\theta = \int_0^{\infty} e^{-x^2 y^2} y\, dx$$

or

$$2e^{-y^2}\theta = \int_0^{\infty} e^{-y^2(1+x^2)} 2y\, dx.$$

Integrating from 0 to ∞ with respect to y and interchanging the order of integration we obtain

$$2\theta^2 = \int_0^{\infty} \frac{dx}{1 + x^2} = \left. \tan^{-1} x \right]_0^{\infty} = \tfrac{1}{2}\pi,$$

from which it follows that

$$\int_0^{\infty} e^{-x^2}\, dx = \tfrac{1}{2} \sqrt{\pi}.$$

(This derivation also appears in [B44], p. 99.)

5. Applications of Single Integration

I have already advocated a heuristic approach to the setting up of problems so that the student may come to look upon the calculus as a familiar and natural tool to use. Thus I would emphatically eliminate the use of formulas as much as possible. An emphasis on a heuristic approach does not imply an ignoring of the difficulties of theory, however, and I want to see a theorem such as Duhamel's used when needed to justify the results arrived at intuitively. In this connection it should be mentioned that several substitutes to and variations of Duhamel's theorem have been proposed. See for example, [E7] and [H35].

The main use of areas, volumes, arc lengths, and surfaces in applications is in connection with centroids and moments of inertia. Generally speaking, the areas, volumes, arc lengths, and surfaces are of simple types and from the point of view of applications, more than half of the problems usually considered in calculus courses may be regarded as artificially difficult. In [D22], the author advocates the use of the intrinsic geometry of the figures involved rather than an imposed coordinate system to obtain more intuitive solutions. [K2] advocates teaching moment problems in general and then having centroids as first moments and moments of inertia as second moments. It should also be noted that mechanics sometimes uses the product of inertia ($\int\int xy\, dA$). Also in connection with applications we should note that the engineer has at his disposal extensive tables of moments of inertia relative to an axis through the center of gravity of the area or volume. Hence what needs to be stressed from the point of view of the prospective engineer is such theorems as $I_s = I_g + Ac^2$, where I_g is the moment of inertia of the area A with respect to the centroidal axis g, and s is any axis parallel to and at a distance c from g.

The most frequent application of integrals is in connection with work (and the variations of potential and heat). The integrals obtained here, incidentally, are almost always quite simple. Other uses which are occasionally met with are applications to magnetic field intensity, electromotive force, intensity of light, gravitational attraction, variable density for areas, and fluid force.

To sum up: I feel that there is too much emphasis generally given to difficult and artificial problems in connection with applications. This tends to foster the all too prevalent feeling among students that only a "math wizard" can use calculus. Fewer difficult problems and more easy problems, carefully considered in detail, would seem to me a desirable reform.

6. Directional Derivatives and Differentials

Directional derivatives both in the x, y, z directions and in other directions occur rather frequently in electricity, while differentials for more than two variables occur occasionally in mechanics and in physical chemistry.

7. Multiple Integration

The theory of multiple integration is, of course, much too difficult to present to a first year calculus class. At this time, however, the theory of single integration should be reviewed and the theory of multiple integration treated intuitively as a generalization. In [B40] the author notes that $\int_0^1 dx \int_0^x e^{y/x}\, dy$ can be integrated in terms of elementary functions, while $\int_0^1 dy \int_y^1 e^{y/x}\, dx$ cannot.

Applications are not nearly as frequent as for single integration, and those that occur are generally of a simple type. Thus again we find that the problems in textbooks frequently go considerably beyond actual use.

Articles by E. A. Whitman describe the use of models in teaching triple integration [W17] and a film on triple integration [W16].

8. Series

Ample additional material on series can be found in any one of the several books on the subject. Our references here are confined to those dealing with basic or unusual material.

The three commonly used tests in beginning calculus for the testing of the convergence of a series are the comparison test, the ratio test, and the integral test. The first is generalized in [S9] as follows:

Let (1) $u_1 + u_2 + u_3 + \cdots$ and (2) $v_1 + v_2 + v_3 + \cdots$ be two series of positive terms. Then if (3) $\lim_{n \to \infty} u_n/v_n = 1$, the series converge or diverge together. The proof is simple. Because of (3), there exists an N such that $n > N$ implies $|u_n/v_n - 1| < \frac{1}{2}$, and therefore $\frac{1}{2} < u_n/v_n < \frac{3}{2}$. Thus (4) $\frac{1}{2}v_n < u_n < \frac{3}{2}v_n$. If (2) converges, use the second of the inequalities (4) and compare (1) with $\frac{3}{2} \Sigma v_n$. If (2) diverges, use the first of the inequalities, and compare (1) with $\frac{1}{2} \Sigma v_n$. (The essential item here is that there exist c and C such that $c < \dfrac{u_n}{v_n} < C$ for n sufficiently large.)

Also in connection with comparison tests see [I1]. Students are always amazed to learn that the harmonic series diverges, and they will be interested to know that "If we strike out from the harmonic series those terms whose denominators contain the digit 9 at least a times and, *at the same time*, the digit 8 at least b times, the digit 7 at least c times and so on, to the digit 0 at least j times ($a, b, c, \ldots, j$ being any given positive integers), the series so obtained will converge."

The ratio test has been generalized in [D21] as follows:

A series $a_1 + a_2 + a_3 + \cdots$ of real numbers is convergent if $a_n \to 0$ and if the ratios a_{n+1}/a_n satisfy the inequality $-1 \leqq a_{i+1}/a_i \leqq r$, where $0 < r < 1$. (The proof is within the range of comprehension of the better students.)

Finally, there is a generalization of the integral test (due to G. H. Hardy) in [W6] as follows:

Let $f(t)$ be a complex valued function of the real variable t in the interval $1 \leqq t < \infty$, such that $f'(t)$ exists and is integrable to $f(t)$ over any finite interval $1 \leq t \leq T$.   Then if $\int_1^\infty f'(t)\,dt$ is absolutely convergent, the series $\Sigma f(n)$ and the integral $\int_1^\infty f(t)\,dt$ converge and diverge together.

This theorem, of course, is too advanced for any but the better student.   The proof involves Abel's partial summation formula.

Under the heading of Taylor's theorem we mention first [N3]. The author expresses a dislike for the customary derivation in which an arbitrary function is drawn out of the air and he develops a formula for the increment in the function as x changes from a to x.   He thus obtains, by successive steps, $f(x) = f(a) + f'(a) \int_a^x dx + f''(a) \int_a^x \int_a^x (dx)^2 + \cdots$ which yields Taylor's formula with remainder.

Along somewhat similar lines is [S20].   Here a direct proof of Taylor's theorem which does not require a knowledge of infinite series or convergence is given.   It is, however, less motivated than the proof in [N3].   See also [N4] and [S33].   The latter article shows how L'Hospital's rule can be used to determine the coefficients in the expansion of $f(x)$.   For example, if $\sin x = a_0 + a_1 x + a_2 x^2 + \cdots$, if $x = 0$, $a_0 = 0$.   Thus

$$\frac{\sin x}{x} = a_1 + a_2 x + a_3 x^2 + \cdots.$$

and $\qquad \underset{x \to 0}{L}\ (\sin x/x) = 1 = a_1$ etc.

Finally we mention an article, [H32], giving a generalization of Taylor's expansion.   The series developed is given by

$$f(x) = f(a) + \sum_{k=1}^{s} \frac{(m + n - k)!}{(m + n)!} [{}_mC_k f^{(k)}(a)$$
$$- (-1)^k {}_nC_k f^{(k)}(x)](x - a)^k + R$$

where

$$R = (-1)^n \frac{m!\,n!(x - a)^{m+n+1}}{(m + n)!(m + n + 1)!} f^{(m+n+1)}(\theta),\ 0 < \theta < x,$$

and it converges about twice as rapidly as the Taylor expansion (if $n = 0$ we have Taylor's series). The proof requires a knowledge of the formula for repeated integration by parts and of beta functions.

The Taylor's series expansions for sine, cosine, and logarithm occur fairly frequently in physics. Other series appear mainly in connection with the solution of differential equations. For term-wise differentiation of power series see [A9] which uses only the mean-value theorem of differential calculus.

Additional references for this chapter are [B36], [F7], [G4], [H36], [J6], and [R20].

Bibliography

(AMM = American Mathematical Monthly, SSM = School Science and Mathematics, MM = Mathematics Monthly or National Mathematics Magazine prior to 1946, MT = Mathematics Teacher)

A1 Aklar, A. "On the Definition of the Riemann Integral." *AMM*. 67(1960):897–900.

A2 Albert, A. A. "A Suggestion for a Simplified Trigonometry." *AMM*. 50(1943):251–253.

A3 Albert, A. A. *College Algebra*. New York: McGraw-Hill, 1946.

A4 Albert, R. G. "A Paradox Relating to Mathematical Induction." *AMM*. 57(1950):31–37.

A5 Allendoerfer, C. B. "Mathematics for Liberal Arts Students." *AMM*. 54(1947):573–578.

A6 *American Mathematical Monthly*, Index of vol. 1 to 56 inclusive, 57(1950):No. 7, Part II.

A7 Anning, N. "The Pascal Triangle." *AMM*. 37(1930):22–25.

A8 Apostol, T. M. *Calculus*. New York: Blaisdell Publishing Co., 1961.

A9 Apostol, T. M. "Term-Wise Differentiation of Power Series." *AMM*. 59(1952):323–326.

A10 Arnold, W. C. "How to Study Mathematics." *AMM*. 47(1940):704–707.

A11 Ayres, W. L. "Interesting the Engineering Student." *AMM*. 51(1944):200–205.

B1 Bakst, A. *Approximate Computation*. National Council of Teachers of Mathematics, 12th Yearbook. New York: Bureau of Publications, Teachers College, Columbia University, 1937.

B2 Bakst, A. "Conic Sections Formed by Some Elements in a Plane Triangle." *MT*. 24(1931):28–30.

B3 Ballantine, J. P. and Jerbert, A. P. "Distance from a Line, or Plane, to a Point." *AMM*. 59(1952):242–244.

B4   Barbour, M. "Solution to E753." *AMM*. 54(1947):414–416.

B5   Barnett, I. "Proposal for the Improvement of the Teaching of Mathematics." *MM*. 9(1934):74–81.

B6   Barzun, J. *Teacher in America*. Boston: Little, Brown and Co., 1945.

B7   Bateman, R. "Method of Construction of a Parabola." *SSM*. 42(1942):718.

B8   Beatley, R. "The Pupil's Advocate." *AMM*. 57(1950):369–381.

B9   Beberman, M. and Meserve, B. E. "Graphing in Elementary Algebra." *MT*. 49(1956):260–266.

B10 Bell, E. T. "Budda's Advice to Students and Teachers of Mathematics." *MT*. 33(1940):252–261.

B11 Bell, E. T. *Men of Mathematics*. New York: Simon and Schuster, 1937.

B12 Bell, E. T. "On Proof by Mathematical Induction." *AMM*. 27(1920):413–415.

B13 Bell, E. T. *The Development of Mathematics*. New York: McGraw-Hill Book Co., 1940.

B14 Bell, E. T. "The Place of Rigor in Mathematics." *AMM*. 41(1934):599–607.

B15 Bell, E. T. *The Search for Truth*. London: George Allen and Unwin Ltd., 1935.

B16 Benbow, C. "Means and Ends in Mathematics." *AMM*. 49(1942):105–106.

B17 Bennett, A. A. "On the Treatment of Maxima and Minima in Calculus Texts." *AMM*. 31(1924):293–294.

B18 Bennett, A. A. "The College Teacher of Mathematics Looks at Teacher Training." *AMM*. 46(1939):213–223.

B19 Bennett, A. A. "The Consequences of Rolle's Theorem." *AMM*. 31(1924):40–42.

B20 Bennett, A. A. "The Definition of a Radian." *AMM*. 32(1925):509–510.

B21 Bennett, A. A. "The Sign of the Distance in Analytic Geometry." *AMM*. 26(1919):344–350.

B22 Betz, W. *Fifth Yearbook*, National Council of Teachers of Mathematics, pp. 149–198. New York: Bureau of Publications, Teachers College, Columbia University, 1930.

B23 Bingley, G. A. "The Complex Variable in the Solution of Problems in Elementary Analytic Geometry." *AMM*. 33(1926):418–421.

B24 Bird, M. T. "Simplification of Equations of Conics." *AMM*. 54(1947):104–106.

B25 Birkhoff, G. and MacLane, S. *A Survey of Modern Algebra*. New York: The Macmillan Co., 1941.

B26 Birkhoff, G. B. "The Mathematical Nature of Physical Theories." *American Scientist*. 31(1943):282–310.

B27   Black, M. "The Relevance of Mathematical Philosophy to the Teaching of Mathematics." *Mathematical Gazette.* 22(1938): 149–163.

B28   Bliss, G. A. "Mathematical Interpretation of Geometrical and Physical Phenomena." *AMM.* 40(1933):472–480.

B29   Blumberg, H. "On the Change of Form." *AMM.* 53(1946): 181–192.

B30   Borman, J. L. "A Remark on Integration by Parts." *AMM.* 51(1944):32–33.

B31   Bourne, S. "Coordinate Geometry from the Vector Point of View." *AMM.* 59(1952):245–248.

B32   Bowen, R. O. *The New Professors.* New York: Holt, Rinehart and Winston, 1960.

B33   Boyer, C. B. "Cardan and the Pascal Triangle." *AMM.* 57(1950): 387–390.

B34   Boyer, C. B. "Clairaut and the Origin of the Distance Formula." *AMM.* 55(1948):556–557.

B35   Brauer, A. "The Proof of the Law of Sines." *AMM.* 59(1952):319.

B36   Brown, A. B. "To Text-book Writers—and Readers." *AMM.* 47(1940):375–381.

B37   Bruce, R. "Equivalence of Equations in One Unknown," *MT.* 24(1931):238–244.

B38   Bruns, W. "The Introduction of Negative Numbers." *MT.* 33(1940):262–269.

B39   Buchanan, H. E. "A Manual for Young Teachers of Mathematics." *AMM.* 53(1946):371–377.

B40   Buck, R. C. "Brief Notes and Comments #17." *MM.* 20(1945):33.

B41   Burgess, H. T. "Relating to the Derivation of the Distance Formula." *AMM.* 25(1918):181.

B42   Bush, L. E. "The Introduction of Invariant Theory into Elementary Analytic Geometry." *MM.* 12(1937):82–89 and 131–137.

B43   Bussey, W. H. "The Origin of Mathematical Induction." *AMM.* 24(1917):199–207.

B44   Byerly, W. E. *Elements of the Integral Calculus,* 2nd edition, Boston: Ginn and Co., 1889.

C1   Cameron, E. "Some Observations on Undergraduate Mathematics in American Colleges and Universities." *AMM.* 60(1953): 151–155.

C2   Campbell, A. D. "Advice to the Graduate Assistant." *AMM.* 45(1938):32–33.

C3   Campbell, A. D. "A Note on the Function $y = a^x (a < 0)$." *AMM.* 34(1927):203.

C4   Cantor, N. *Dynamics of Learning.* Buffalo, N.Y.: Foster and Stewart, 1946.

C5   Carver, W. B. "Thinking versus Manipulation." *AMM.* 44(1937): 359–363.

C6   Cell, J. W. "Mathematics Instruction in Engineering." *AMM*. 50(1943):303–307.

C7   Cell, J. W. "The Trigonometric Functions." *MT*. 43(1950): 187–192.

C8   Chrystal, G. *Algebra, vol. I*, 3rd edition. London and Edinburgh: A. and C. Black, 1893.

C9   Chu, T. S. "Angles with Rational Tangents." *AMM*. 57(1950): 407–408.

C10 Church, A. "Differentials," *AMM*. 49(1942):389–392.

C11 Cirul, J. W. "A Method for Solving Quadratic Equations." *AMM*. 44(1937):462–463.

C12 Clair, H. S. "The Laws of Algebra and Modern Algebra." *SSM*. 53(1953):29–33.

C13 Cogan, L. J. "A Survey of Courses in Mathematics for the Liberal Arts Student." *AMM*. 62(1955):319–322.

C14 Cole, L. *The Background for College Teaching*. New York: Farrar and Rinehart, 1940.

C15 Collegiate Mathematics Needed in the Social Sciences, *AMM*. 39(1932):569–577.

C16 Committee on the Undergraduate Program in Mathematics. *Course Guides for the Training of Teachers of Junior High and High School Mathematics*. Buffalo, N.Y.: Mathematical Association of America, 1961.

C17 Committee on the Undergraduate Program in Mathematics. *Recommendations for the Training of Teachers of Mathematics*. Buffalo, N.Y.: Mathematical Association of America, 1961.

C18 Committee on the Undergraduate Program in Mathematics. *Recommendations on the Undergraduate Mathematics Program for Engineers and Physicists*. Buffalo, N.Y.: Mathematical Association of America, 1962.

C19 Coolidge, J. L. "The Story of the Binomial Theorem." *AMM*. 56(1949):147–157.

C20 Corcoran, G. F. "Mathematics and Electrical Engineering." *Journal of Engineering Education*. 37(1947):818–820.

C21 Corliss, J. I., Feinstein, J. K. and Levin, H. S. *Analytic Geometry*. New York: Harper & Bros., 1949.

C22 Courant, R., and Robbins, H. *What is Mathematics?* London and New York: Oxford University Press, 1941.

C23 Cronkhite, B. B. (editor). *A Handbook for College Teachers*. Cambridge, Mass.: Harvard University Press, 1950.

D1   Dadourian, H. M. *How to Study, How to Solve–Arithmetic-Calculus*. Cambridge, Mass.: Addison-Wesley Press, 1951.

D2   Dadourian, H. M. "How to Make Mathematics More Attractive." *MT*. 53(1960):548–551.

D3   Daniell, P. J. "New Rules of Quadrature." *AMM*. 24(1917): 109–113.

D4 Daniels, F. "Mathematics for Students of Chemistry." *AMM*. 35(1928):3–9.

D5 Daniels, F. and others. "The Mathematical Training of Chemists." *AMM*. 40(1933):1–4.

D6 Davis, D. R. *The Teaching of Mathematics*. Cambridge, Mass.: Addison-Wesley Press, 1951.

D7 Davis, H. T. *College Algebra*. New York: Prentice-Hall, 1942.

D8 Dawson, J. "Cultural Mathematics at Antioch." *School and Society*. 33(1931):141–144.

D9 Dehn, E. "Reduction of Angles in Trigonometry." *MT*. 28(1935): 320–321.

D10 Dexter, L. A. and Thornton, R. A. "On the Analysis of Transfer of Training." *American Journal of Physics*. 19(1951):538–545.

D11 Dewey, J. *How We Think*. Boston: D. C. Heath & Co., 1933.

D12 Dickinson, E. L. and Ruch, G. M. "Analysis of Certain Difficulties in Factoring in Algebra." *Journal of Educational Psychology*. 16(1925):323–328.

D13 Dorwart, H. L. "Beyond Quadratics." *MM*. 16(1941–42): 231–237.

D14 Dorwart, H. L. "Values of the Trigonometric Ratios of $\pi/8$ and $\pi/12$." *AMM*. 49(1942):324–325.

D15 Dresden, A. *"An Invitation to Mathematics*. New York: Henry Holt & Co., 1936.

D16 Dresden, A. "A Program for Mathematics." *AMM*. 42(1935): 198–208.

D17 Dubisch, R. "Impossible and Unsolved Problems in Elementary Mathematics." *MT*. 41(1948):285–286.

D18 Dubisch, R., Howes, V. E., and Bryant, S. J. *Intermediate Algebra*. New York: John Wiley and Sons, 1960.

D19 Dubisch, R. *Trigonometry*. New York: The Ronald Press, 1955.

D20 Dudley, A. "Type of Mathematical Training Needed by Electrical Engineers." *AMM*. 42(1935):301–306.

D21 Duffin, R. J. "A Generalization of the Ratio Test for Series." *AMM*. 55(1948):153–155.

D22 Duncan, D. G. "Some Standard Problems in Integration Simplified." *AMM*. 61(1954):421–422.

D23 Dunkel, O. "Relating to the Exponential Function." *AMM*. 24(1917):244–246.

D24 Dwight, H. B. *Tables of Integrals and Other Mathematical Data*. New York: The Macmillan Co., 1947.

E1 Eaton, T. H. *College Teaching*. New York: John Wiley and Sons, 1932 (out of print).

E2 Eaves, J. C. "Off the Beaten Path with Some Differentiation Formulas for the Trigonometric, Exponential, and Logarithmic Functions." *MM*. 26(1953):147–152.

E3 *Education Index*. New York: H. Wilson & Co. vols. 1–19. 1929–1962.

E4   Edwards, W. H. "Trigonometric Formulae Encountered in a College Engineering Course." *SSM.* 28 (1928):239–243.

E5   Ellis, C. "Mathematics and Engineering Education." *SSM.* 35(1935):123–132.

E6   Erskine, W. H. "The Content of a Course in Algebra for Prospective Secondary School Teachers of Mathematics." *AMM.* 46(1939): 32–35.

E7   Ettlinger, H. J. "A Simple Form of Duhamel's Theorem and Some New Applications." *AMM.* 29(1922):239–250.

E8   Evans, J. P. "The Extended Law of the Mean by a Translation-Rotation of Axes." *AMM.* 67(1960):580–581.

E9   Eves, H. "The Names 'Ellipse,' 'Parabola,' and 'Hyperbola'." *MT.* 53(1960):280–281.

E10   Eves, H. "The Irrationality of $\sqrt{2}$." *MT.* 38(1945):317–318.

E11   Eves, H., and Newsom, C. V. *An Introduction to the Foundations and Fundamental Concepts of Mathematics.* New York: Rinehart & Co., 1958.

F1   Farrell, A. B. "On the Second Derivative." *AMM.* 52(1945):207.

F2   Fehr, H. F. *Secondary Mathematics.* Boston: D. C. Heath & Co., 1951.

F3   Fehr, H. F. "The Quadratic Equation." *MT.* 26(1933):146–149.

F4   Fertig, R. "Substitute for the Law of Tangents." *AMM.* 48(1941): 132.

F5   Fettis, H. E. "On Various Methods of Solving Cubic Equations." *MM.* 17(1942–43):117–129.

F6   Fine, H. B. *The Numbers System of Algebra Treated Theoretically and Historically*, Reprint from 1890 edition. New York: G. E. Stechert & Co., 1937.

F7   Ford, W. B. "The Teaching of the Calculus." *AMM.* 17(1910): 77–81.

F8   Fort, M. K., Jr. "The Maximum Value for a Continuous Function." *AMM.* 58(1951):32–33.

F9   Fort, M. K. "Differentials." *AMM.* 59(1952):392–395.

F10   Fort, T. *Calculus.* Boston: D. C. Heath & Co., 1951.

F11   Fulton, C. M. "A Simple Proof of the Binomial Theorem." *AMM.* 59(1952):243.

G1   Garve, R. "A Note on the Function $y = x^x$." *AMM.* 34(1927): 429.

G2   Gerhard, F. B. "The Actuarial Profession." *AMM.* 55(1948): 481–487.

G3   Good, C. V. *Teaching in College and University.* Baltimore: Warwick and York, 1929.

G4   Graham, B. "Some Calculus Suggestions by a Student." *AMM.* 24(1917):265–271.

G5   Griffin, C. W. "Significant Figures." *MM.* 10(1935):20–24.

G6 Griffin, F. L. "The Undergraduate Mathematical Curriculum of the Liberal Arts College." *AMM*. 37(1930):46–54.

G7 Griffin, F. L. "Undergraduate Mathematical Research." *AMM*. 49(1942):379–385.

G8 Grintner, L. E. "Teaching Mathematics to Engineers." *Journal of Engineering Education*. 37(1947):536–539.

H1 Haag, V. E. *Structure of Elementary Algebra*, vol. III of *Studies in Mathematics* of the School Mathematics Study Group. A. C. Vroman, Pasadena, Calif., 1961.

H2 Hadamard, J. *The Psychology of Invention in the Mathematical Field*. Princeton, N.J.: Princeton University Press, 1949.

H3 Hamilton, H. J. "Toward Understanding Differentials." *AMM*. 59(1952):398–403.

H4 Hamilton, N. T., and Landin, J. *Set Theory: The Structure of Arithmetic*. Boston: Allyn and Bacon, 1961.

H5 Hammer, P. C. "Mantissa and Characteristic." *AMM*. 49(1942): 245–246.

H6 Hamming, R. W. "The Transcendental Nature of cos X." *AMM*. 52(1945):336–337.

H7 Harper, F. S. "A Method for the Point by Point Construction of Central Conics by Ruler and Compass." *MM*. 21(1947–48):55–57.

H8 Harris, J. A. "The Fundamental Mathematical Requirements of Biology." *AMM*. 36(1929):179–198.

H9 Hart, J. J. "A Correction for the Trapezoidal Rule." *AMM*. 59(1952):33–37. (Correction to the article given on p. 406.)

H10 Hassler, J. O. and others. "Readings in the Literature on Teaching with Special Reference to Mathematics." *AMM*. 42(1935): 472–476.

H11 Hawthorne, F. "Derivation of the Equation of Conics." *AMM*. 54(1947):219–220.

H12 Hayes, J. M. "Mathematics in Civil Engineering." *Journal of Engineering Education*. 38(1947):220–222.

H13 Hedrick, E. R. "Desirable Cooperation between Educationalists and Mathematicians." *School and Society*. 36(1932):769–777.

H14 Hedrick, E. R. and others. "Four Papers on the Teaching of Mathematics." *Bull. 19, Society for the Promotion of Engineering Education*, 1932.

H15 Hedrick, E. R. "Teaching for Transfer of Training in Mathematics." *MT*. 30(1937):51–55.

H16 Hedrick, E. R. "Teaching of Trigonometry for Engineers." *Journal of Engineering Education*. 22(1931):299–310.

H17 Helsel, R. G. and Rado, T. "Can We Teach Good Mathematics to Undergraduates?" *AMM*. 55(1948):28–29.

H18 Hempel, C. G. "On the Nature of Mathematical Truth." *AMM*. 52(1945):543–556.

H19 Henderson, K. B. and Dickman, K. "Minimum Mathematical Needs of Prospective Students in a College of Engineering." *MT*. 45(1952):89–93.

H20 Henkin, L. *Mathematical Induction, MAA Film Manual* No. 1. Buffalo, N.Y.: Mathematical Association of America, 1961.

H20a Henkin, L. "On Mathematical Induction." *AMM*. 67(1960): 323–338.

H21 Higgins, T. J. "Power Series Expansions of Transcendental Functions." *SSM*. 40(1940):265–266.

H22 Highet, G. *The Art of Teaching*. New York: Alfred A. Knopf, 1950.

H23 Hildebrandt, T. H. "Relating to Finding Derivatives of Trigonometric Functions." *AMM*. 25(1918):125–126.

H24 Hoer, R. S. "On Proofs by Mathematical Induction." *AMM*. 29(1922):162–164.

H25 Hoffman, S. "A Classroom Proof of $\lim\limits_{t\to0} \dfrac{\sin t}{t} = 1$." *AMM*. 67(1960):671–672.

H26 Hohn, E. "An Existence Proof for Logarithms." *AMM*. 50 (1943):165–166.

H27 Hohn, T. E. "The Conference on Training in Applied Mathematics." *AMM*. 61(1954):242–244.

H28 Householder, A. S. "The Addition Formulas in Trigonometry." *AMM*. 54(1947):326–327.

H29 Howland, W. "Mathematics in Civil Engineering." *SSM*. 35(1935):351–360.

H30 Huff, G. B. "On Defining Conic Sections." *AMM*. 62(1955): 250–251.

H31 Hull, R. "The Mathematical Training of Engineers." *AMM*. 60(1953):106.

H32 Hummel, P., and Seebeck, C. L., Jr. "A Generalization of Taylor's Expansion." *AMM*. 56(1949):243–247.

H33 Hummel, P. M. "A Note on Interpolation." *AMM*. 54(1947): 218–219.

H34 Huntington, E. V. "An Elementary Theory of the Exponential and Logarithmic Function." *AMM*. 23(1916):241–246.

H35 Huntington, E. V. "On Setting up a Definite Integral without the Use of Duhamel's Theorem." *AMM*. 24(1917):271–275.

H36 Huntington, E. V. "Teaching the Calculus." *Journal of Engineering Education*. 22(1932):485–521, 708–710.

I1 Irwin, F. "A Curious Convergent Series." *AMM*. 23(1916): 149–152.

I2 Ivanoff, V. F. "The nth Derivative of a Fractional Function." *AMM*. 55(1948):491.

J1 Jackson, D. "A Comment on 'Differentials'." *AMM*. 49(1942): 389.

J2   James, G. "Remarks on the Definition of Continuity." *AMM.* 44(1937):235.

J3   James, G. "The Cause and Cure of Delinquency in College Mathematics." *MM.* 11(1936–37):247–248.

J4   Jeffery, R. L. "Productive Scholarship in the Undergraduate College." *AMM.* 42(1935):364–369.

J5   Johns, A. E. "The Reduced Equation of a General Conic." *AMM.* 55(1948):100–104.

J6   Johnson, C. A. "Coordinating Calculus Instruction in the Engineering Program." *Journal of Engineering Education.* 41(1951): 537–541.

J7   Johnston, F. S. "Simple Constructions for the Conics." *AMM.* 44(1937):524–525.

J8   Jones, B. W. *Elementary Concepts of Mathematics.* New York: The Macmillan Co., 1947.

J9   Jones, B. W. "On the Training of Teachers for Secondary Schools." *AMM.* 46(1939):428–434.

J10   Jones, P. S. "The Teaching Fellow Program at Michigan." *AMM.* 55(1948):145–147.

K1   Kac, M. and Randolph, J. F. "Differentials." *AMM.* 49(1942): 110–112.

K2   Kaltenborn, H. S. "Teaching Centroids and Moments of Inertia Simultaneously." *MM.* 16(1941–42):299–304.

K3   Karelitz, G. "Mathematics in Mechanical Engineering." *MT.* 32(1939):87.

K4   Karpinski, L. C. "Simultaneous Quadratics Solvable in Quadratic Irrationalities." *AMM.* 43(1936):362–366.

K5   Karst, O. J. "Analysis of the Required Undergraduate Mathematics Courses for Engineering Students." *SSM.* 55(1955):29–39.

K6   Karst, O. J. "The Limit." *MT.* 51(1958):43–49.

K7   Kelley, J. L. *Introduction to Modern Algebra.* Princeton, N.J.: D. Van Nostrand Co., Inc., 1961.

K8   Kells, L. M., and Stotz, H. C. *Analytic Geometry.* New York: Prentice-Hall, 1949.

K9   Kelly, F. J. (editor). *Improving College Instruction,* Series I, Reports of Committees and Conferences No. 48, vol. XV. Washington, D.C.: American Council on Education, 1951.

K10   Kelly, F. J. "Recent Publications in College Teaching—A Brief Review." *Higher Education.* 7(1951):221–225.

K11   Kelly, F. J. *Toward Better Teaching.* Office of Education Bulletin No. 13. Washington, D.C.: U.S. Government Printing Office, 1950.

K12   Kelly, L. M. "Problem E753." *AMM.* 54(1947):38.

K13   Kemeny, J. G. "Rigor vs. Intuition in Mathematics." *MT.* 54(1961):66–74.

K14   Kempner, A. J. "Need for Cooperation between High School and College Teachers of Mathematics." *MT.* 31(1938):117–123.

K15   Kempner, A. J. "The Deadly Parallel." *AMM*. 46(1939):280.

K16   Kennedy, E. C. "A Note on a Certain Trigonometric Equation."
MM. 11(1936–37):322.

K17   Kholodovsky, E. A. "Graphical Integration and Differentiation
of Functions in a Polar Coordinate System." *AMM*. 36(1929):3–21.

K18   Kinney, L. B., and Purdy, C. R. *Teaching Mathematics in the
Secondary Schools*. New York: Rinehart and Co., 1952.

K19   Klein, F. *Elementary Mathematics from an Advanced Standpoint*,
Translated by E. R. Hedrick and C. A. Noble from the 3rd German
ed., 2 vols. New York: Dover Publications, 1945.

K19   Knebelman, M. S. "An Elementary Limit." *AMM*. 50(1943):507.

K20   Krathwohl, W. C. "Some Effective Methods of Teaching Mathematics." *Journal of Engineering Education*. 39(1948):81–85.

L1   Landau, E. G. H. *Foundations of Analysis*, translated by F. Stein-
hardt. New York: Chelsea Publishing Co., 1951.

L2   Lange, L. "On a Three-dimensional Presentation of Functions of a
Complex Variable." *AMM*. 46(1939):190–198.

L3   Lange, L. "Still Another Viewpoint." *SSM*. 41(1941):860–869.

L4   Lange, L. "The Law $\sqrt{a}\,\sqrt{b} = \sqrt{ab}$ Holds for All Numbers."
SSM. 41(1941):457–461.

L5   Lange, L. "Deriving the Equations of the Sections of a Cone."
AMM. 63(1956):488–491.

L6   Lazar, N. "One Unknown of Two?" *MT*. 26(1933):176–182.

L7   Leacock, S. "Human Interest Put into Mathematics." *MT*.
22(1929):302–304, or *Literary Lapses*. New York: Dodd, Mead and
Co., 1937.

L8   Lefschetz, S. "A Direct Proof of De Moivre's Theorem." *AMM*.
23(1916):366–368.

L9   Levi, H. *Elements of Algebra*. New York: Chelsea Publishing Co.,
1954.

L10   Levy, H. "Analytic Geometry and the Calculus." *AMM*.
68(1961):925–927.

L11   Ligda, P. "Textbook Solutions of Algebraic Problems." *SSM*.
27(1927):41–45.

L12   Ligda, P. *The Teaching of Elementary Algebra*. Boston: Houghton-
Mifflin Co., 1925.

L13   Liming, R. A. *Practical Analytic Geometry with Applications to
Aircraft*. New York: The Macmillan Co., 1944.

L14   Lippitt, V. G. "The Teaching of Mathematics as Viewed by an
Engineer." *SSM*. 44(1944):505–510.

L15   Longley, W. "First Exercises in Differentials." *MM*. 10(1936):
219–226.

L16   Lovitt, W. V. "Inverse Trigonometric Functions." *AMM*.
27(1920):117–119.

L17   Lyon, R., and Ward, M. "The Limit for *e*." *AMM*. 59(1952):
102–103.

M1   MacDuffee, C. C. "Objectives in Calculus." *AMM*. 54(1947): 335–337.

M2   Mancill, J. D. "One-sided Maxima and Minima." *AMM*. 55(1948):311–315.

M3   Manheim, J. "Word Problems or Problems with Words." *MT*. 54(1961):234–238.

M4   Mansfield, R. "The General Equation of the Conic." *SSM*. 42(1942):729–736.

M5   Mathematics Division of the Society for the Promotion of Engineering Education. *Engineering Problems Illustrating Mathematics*. New York: McGraw-Hill Book Co., 1943.

M6   Mathews, R. M. "Relating to the Derivation of a Distance Formula." *AMM*. 24(1917):476–477.

M7   Mathews, R. M. "The Law of Cosines versus the Law of Tangents." *SSM*. 16(1916):41–45.

M8   May, K. "The Straight Line Treated by Translation." *MM*. 22(1948–49):211

M9   Mazkewitsch, D. "Coefficients in Expansions of Polynomials." *SSM*. 58(1958):703–709.

M10   McKeachie, W., with Kimble, G. *Teaching Tips: A Guide-Book for the Beginning College Teacher*. 4th ed. Ann Arbor, Mich.: The George Wahr Publishing Co., 1960.

M11   McShane, E. J. "The Addition Formulas of Elementary Trigonometry." *AMM*. 48(1941):688–689.

M12   Meserve, B. E., and Sobel, M. A., *Mathematics for Secondary School Teachers*. Englewood Cliffs, N.J.: Prentice-Hall, 1962.

M13   Miller, F. H. "Artificial Mathematics." *AMM*. 57(1950):329-330.

M14   Miller, G. A. "A Crusade Against the Use of Negative Numbers." *SSM*. 33(1933):955–964.

M15   Miller, G. A. "Reduction of the Trigonometric Functions of any Angle to the Functions of the Angles in a Small Interval." *AMM*. 18(1911):171–174.

M16   Miller, G. H. "Difficulties in Algebra: A Study." *SSM*. 58(1958): 714–719.

M17   Millett, T. B. Professor: Problems and Rewards of College Teaching. New York: The Macmillan Co., 1961.

M18   Minarik, R. "Are We Teaching Engineering Mathematics?" *Journal of Engineering Education*. 25(1935):499–509.

M19   Miser, W. L. "Summing Series Whose General Terms are Polynomials." *MM*. 7(1931–32):17–19.

M20   Moritz, R. E. "On the Geometrical Representation of Indeterminate Forms." *AMM*. 28(1921):211–217.

M21   Morris, F. C. *Effective Teaching*. New York: McGraw-Hill Book Co., 1950.

M22   Muir, T. *The Theory of Determinants in the Historical Order of Development*, 4 vols. London: Macmillan and Co., Ltd., 1906–23.

M23 Munro, W. B., and others. "Report of the Committee on College and University Teaching." *Bulletin of the American Association of University Professors.* 29(1933):sec. 2, 1–122.

M24 Munro, W. D. "Note on the Euler-MacLaurin Formula." *AMM.* 65(1958):201–203.

M25 Munroe, M. E. "Bringing Calculus Up-to-Date." *AMM.* 65(1958): 81–90.

M26 Murnaghan, F. D. *Analytic Geometry.* New York: Prentice-Hall, 1946.

M27 Murnaghan, F. D. "The Teaching of College Mathematics." *AMM.* 53(1946):419–425.

N1 Newsom, C. V., and Randolph, J. F. "Trigonometry Without Angles." *MT.* 39(1946):66–70.

N2 Newsom, C. V. "The Teaching of Collegiate Mathematics." *SSM.* 52(1952):130–132.

N3 Nicholas, C. P. "Taylor's Theorem in a First Course." *AMM.* 58(1951):559–561.

N4 Nichols, C. P. "More on Taylor's Theorem in a First Course." *AMM.* 60(1953):329.

N5 Nichols, E. D., and Garland, H. *The Teaching of Contemporary Secondary Mathematics.* New York: Holt, Rinehart and Winston, 1963.

N6 Nickle, G. H. "Mathematics Most Used in the Sciences of Physics, Chemistry, Engineering and Higher Mathematics." *MT.* 35(1942): 77–83.

N7 Northrop, E. P. "Mathematics in a Liberal Education." *AMM.* 52(1945):132–137.

N8 Northrop, E. P. "The Mathematics Program at the College of the University of Chicago." *AMM.* 55(1948):1–7.

N9 Nowlan, F. S. "Objectives in the Teaching of College Mathematics." *AMM.* 57(1950):73–82.

N10 Nyberg, J. "Methods of Teaching Verbal Problems." *National Council of Teachers of Mathematics, 7th Yearbook,* pp. 167–179. New York: Bureau of Publications, Teachers College, Columbia University, 1932.

N11 Nyberg, J. A. "A Consistent Measurement of Lines and Angles." *SSM.* 37(1937):330–333.

N12 Nyberg, J. A. "Notes from a Mathematics Classroom." *SSM.* 42(1942):565–568.

O1 Oakley, C. O. "End-point Maxima and Minima." *AMM.* 54(1947):407–409.

O2 O'Brien, K. E. "Problem Solving." *MT.* 49 (1956):79–86.

O3 Ogilvy, C. S. "The Transition from School to College Mathematics." *MT.* 53(1960):514–517.

O4 Ogilvy, C. S. "Mathematics Vocabulary of Beginning Freshmen." *AMM.* 56(1949):261–262.

O5   Olds, E. G.   "Remarks on the Law of Cosines."   *MM*. 11(1936–37): 324–326.

O6   Olmstead, J. M. H.   "Rational Values of Trigonometric Functions." *AMM*. 52(1945):507–508.

O7   Ore, O.   "Mathematics for Students of the Humanities."   *AMM*. 51(1944):453–458.

O8   Osborn, R.   "A Note on Indeterminate Forms."   *AMM*. 59(1952): 549.

P1   Palmer, C. I. and Leigh, C. W.   *Plane and Spherical Trigonometry.* New York: McGraw-Hill Book Co., 1934.

P2   Patterson, B. C.   "The Components for Velocity and Acceleration." *AMM*. 42(1935):554–557.

P3   Pease, E. M. J.   *Intermediate Algebra.*   New York: Prentice-Hall, 1950.

P4   Peirce, B. O.   *A Short Table of Integrals.*   Boston: Ginn and Co., 1929.

P5   Phipps, C. G.   "The Relation of Differentials and Delta Increments."   *AMM*. 59(1952):395–398.

P6   Polya, G.   *How to Solve It.*   Princeton, N.J.: Princeton University Press, 1945.

P7   Polya, G.   "Generalization, Specialization, Analogy."   *AMM*. 55(1948):241–243.

P8   Polya, G.   "With, or Without, Motivation?"   *AMM*. 56(1949): 684–691.

P9   Polya, G.   *Induction and Analogy in Mathematics.*   Princeton, N.J.: Princeton University Press, 1954.

P10 Polya, G.   *Mathematical Discovery.*   New York: John Wiley and Sons, 1962.

P11 Polya, G.   *Patterns of Plausible Inference.*   Princeton, N.J.: Princeton University Press, 1954.

P12 Porges, A.   "Again That Quadratic Equation."   *SSM*. 44(1944): 565–568.

P13 Porter, W. B.   "The Derivative of the Logarithm."   *AMM*. 23(1916):204–206.

P14 Prenowitz, W.   "Insight and Understanding in the Calculus." *AMM*. 60(1953):32–37.

P15 Price, G. B.   "Some Identities in the Theory of Determinants." *AMM*. 54(1947):75–89.

P16 Putnam, T. M.   "Mathematical Forms of Certain Eroded Mountain Sides."   *AMM*. 24(1917):451–453.

P17 Putney, T.   "Proof of the First Mean Value Theorem of the Integral Calculus."   *AMM*. 60(1953):113.

Q1   Querry, J. W.   "A Survey Course for Teachers."   *AMM*. 50(1943): 176–178.

R1   Randolph, J. F. and Kac, M.   *Analytic Geometry and Calculus.* New York: The Macmillan Co., 1947.

R2   Ransom, W. R.   "A Geometrical Meaning for $f''(x)$."   *SSM*. 43(1943):600.

R3   Ransom, W. R.   "Bringing in Differentials Earlier."   *AMM*. 58(1951):336–337.

R4   Ransom, W. R.   "Introducing $e = 2.718+$."   *AMM*. 55(1948):572.

R5   Rasey, M. I.   *This Is Teaching*.   New York: Harper & Bros., 1950.

R6   Raynor, G. E.   "Mathematical Induction."   *AMM*. 33(1926): 376–377.

R7   Read, C. B. and Klein, A. E.   "An Analysis of Mathematics Courses in Four Year Colleges."   *SSM*. 55(1955):40–55.

R8   Read, C. B.   "Is a Mantissa Necessarily Positive?"   *AMM*. 48(1941):203–204.

R9   Read, C. B.   "Logarithms vs. Cologarithms."   *SSM*. 36(1936): 981–985.

R10   Read, C. B.   "Principal Values of Inverse Trigonometric Functions."   *SSM*. 40(1940):343–344.

R11   Read, C. B. and Hitt, J. K.   "Undefined Expressions · Involving Fractional Exponents."   *SSM*. 39(1939):839.

R12   Reeve, W. D.   *Mathematics for the Secondary School*.   New York: Henry Holt and Co., 1954

R13   Reichenbach, H.   *The Rise of Scientific Philosophy*.   Berkeley and Los Angeles: University of California Press, 1941.

R14   Richards, O. W.   "The Mathematics of Biology."   *AMM*. 32(1925): 30–36.

R15   Richardson, M.   *Fundamentals of Mathematics*.   New York: The Macmillan Co., 1947.

R16   Richardson, M.   "Mathematics and Intellectual Honesty."   *AMM*. 59(1952):73–78.

R17   Richardson, M.   "On the Teaching of Elementary Mathematics."   *AMM*. 49(1942):498–505.

R18   Richmond, D. E.   *Fundamentals of the Calculus*.   New York: McGraw-Hill Book Co., 1950.

R19   Rickart, F. A.   "A Problem in Minimum Values."   *MM*. 13(1937–38): 362–366.

R20   Rietz, H. L.   "On the Teaching of the First Course in Calculus."   *AMM*. 26(1919):341–344.

R21   Rietz, H. L.   "The Teaching of College Algebra."   *AMM*. 17(1910): 51–55.

R22   Riley, J. W., Jr., Ryan, B. F. and Lifschitz, M.   *The Student Looks at His Teacher*.   New Brunswick, N.J.: Rutgers University Press, 1950.

R23   Robbins, H. E.   "A Note on the Riemann Integral."   *AMM*. 50(1943):617–618.

R24   Robinson, L. V.   "Pascal's Triangle and Negative Exponents."   *AMM*. 54(1947):540–541.

R25   Robinson, R. M.   "A Note on Linear Equations."   *AMM*. 56(1949): 251.

R26 Rogers, H., Jr. "A General Education Course in Pure Mathematics." *AMM*. 63(1956):460–465.

R27 Rosenbaum, R. A. "A Note on Joint Variation." *AMM*. 49(1942): 537–538.

R28 Rudman, B. "Teaching the Verbal Problem in Intermediate Algebra." *MT*. 22(1929):83–92.

R29 Rush, M. J. "Some Formulas of Elementary Trigonometry." *AMM*. 28(1921):443–446.

R30 Russell, B. *Introduction to Mathematical Philosophy*. New York: The Macmillan Co., 1920.

S1 Sandham, H. F. "An Approximate Construction for *e*." *AMM*. 54(1947):215–216.

S2 Sandham, H. F. "A Well-known Integral." *AMM*. 53(1946):587.

S3 Sandiford, P. "Transfer of Training." *The School*. 27(1938): 93–97.

S4 Sanford, N., editor. *The American College*. New York: John Wiley and Sons, 1962.

S5 Sanford, V. *History and Significance of Problems in Algebra*. Contributions to Education #251. New York: Bureau of Publications, Teachers College, Columbia University, 1927.

S6 Scarborough, J. S. "Formulas for the Error in Simpson's Rule." *AMM*. 33(1926):76–83.

S7 Schaaf, W. L. *A Bibliography of Mathematical Education*. Forest Hills, N.Y.: Stevinus Press, 1941.

S8 Schaaf, W. L. "Required Mathematics in a Liberal Arts College." *AMM*. 44(1937):445–453.

S9 Scheffe, H. A. "A Series Comparison Test for Calculus Students." *AMM*. 48(1941):256–257.

S10 Schelkunoff, S. A. "A Note on Geometrical Applications of Complex Numbers." *AMM*. 37(1930):301–303.

S11 Schelkunoff, S. A. "Complex Numbers in Elementary Mathematics." *SSM*. 32(1932):284–301.

S12 Scherling, M. G. "A Fraction Rule in Logarithms." *MM*. 16(1936–37):195.

S13 School Mathematics Study Group. *Elementary Functions*. New Haven, Conn.: Yale University Press, 1960.

S14 School Mathematics Study Group. *First Course in Algebra*. New Haven, Conn.: Yale University Press, 1960 (also teacher's commentary).

S15 School Mathematics Study Group. *First Course in Geometry*. New Haven, Conn.: Yale University Press, 1960.

S16 School Mathematics Study Group. *Intermediate Mathematics*. New Haven, Conn.: Yale University Press, 1960.

S17 School Mathematics Study Group. *Intermediate Mathematics*. Supplementary Unit I. (The Development of the Real Number System). New Haven, Conn.: Yale University Press, 1960.

S18 Schorling, R. "Place of Mathematics in a General Education."
SSM. 40(1940):14–26.

S19 Seebeck, C. L., Jr., and Jewett, J. W. "A Development of Logarithms Using the Function Concept." *AMM*. 64(1957):667–668.

S20 Seebeck, C. L., Jr. "A Note on Taylor's Theorem." *AMM*. 57(1950):32–33.

S21 Seebeck, C. L., Jr., and Jewett, J. W. "More about Logarithms." *AMM*. 65(1958):697–698.

S22 Seidlin, J. A. *A Critical Study of the Teaching of Elementary College Mathematics*. New York: Bureau of Publications, Teachers College, Columbia University, 1931.

S23 Shaw, A. A. "Geometric Applications of Complex Numbers." *SSM*. 31(1931):754–761.

S24 Shaw, J. B. "Imaginary Orders." *MM*. 12(1938–39):63–76.

S25 Slaught, H. E. "The Teaching of Mathematics." *AMM*. 22(1915): 289–292.

S26 Slaught, H. E. "The Teaching of Mathematics in the Colleges." *AMM*. 16(1909):173–177.

S27 Smail, L. "Some Geometrical Applications of Complex Numbers." *AMM*. 36(1929):504–511.

S28 Smith, C. D. "On the Algebra of Mixtures." *MM*. 9(1934–35): 138–141.

S29 Smith, D. E. "On the Origin of Certain Typical Word Problems." *AMM*. 24(1917):64–71.

S30 Smith, V. G. "Engineering Mathematics." *Journal of Engineering Education*. 39(1947):308–312.

S31 *Source Book of Mathematics Applications, National Council of Teachers of Mathematics* 17th Yearbook. New York: Bureau of Publications, Teachers College, Columbia University, 1942.

S32 Spaulding, F. T. "Three Lectures on Learning and Teaching." Bulletin No. 14. *Society for the Promotion of Engineering Education*, 1931.

S33 Spiegel, M. R. "L'Hospital's Rule and Expression of Functions in Power Series," *AMM*. 62(1955):358–360.

S34 Spiegel, M. R. "Mean Value Theorems and Taylor Series." *MM*. 29(1956):263–266.

S35 Spiegel, M. R. "Partial Fractions with Repeated Linear or Quadratic Factors." *AMM*. 57(1950):180–181.

S36 Spiegel, M. R. "Student's Choice in Mathematics—Fundamental Reasoning or Blind Subservience to Rules." *AMM*. 59(1952): 99–100.

S37 Sprague, A. H. *Calculus*. New York: The Ronald Press Co., 1952.

S38 Stabler, E. R. "Another Viewpoint in Multiplication of Radicals." *SSM*. 41(1941):855–860.

S39 Stabler, E. R. "Demonstrative Algebra." *MT*. 39(1946):255–260.

S40 Stalley, R. "A Generalization of the Geometric Sum." *AMM*. 56(1949):325–327.

S41   Steen, F. H. "Simplification by Rotation." *MM*. 15(1940–41): 369–374.

S42   Stewart, L. "The Binomial Theorem." *MT*. 53(1960):344–348.

S43   Sutton, R. M. "An Instrument for Drawing Confocal Ellipses and Hyperbolas." *AMM*. 50(1943):253–254.

 T1   Tarski, A. *Introduction to Logic*. Translated by O. Helmer. New York: Oxford University Press, 1941.

 T2   Taylor, A. E. *Calculus with Analytic Geometry*. Englewood Cliffs. N.J.: Prentice-Hall, Inc., 1959.

 T3   Taylor, A. E. "Derivatives in the Calculus." *AMM*. 49(1942): 631–642.

 T4   Taylor, A. E. "L'Hospital's Rule." *AMM*. 59(1952):20–24.

 T5   Thomas, G. B., Jr. *Calculus*. Reading, Mass.: Addison-Wesley Publishing Co., 1953.

 T6   Thorndike, E. "Psychology of Problem Solving." *MT*. 15(1922): 212–227, 253–264.

 T7   Tracey, J. "Undergraduate Instruction in Mathematics." *AMM*. 44(1937):284–288.

 T8   Tremblay, A. "Generalization of Pascal's Arithmetical Triangle." *MM*. 11(1936–37):235–238.

 T9   Trine, F. D. "An Introduction to Algebra with Inequalities." *MT*. 53(1960):42–45.

T10   Tukey, J. W. "The Teaching of Concrete Mathematics." *AMM*. 65(1958):1–9.

 V1   Vance, E. P. "Teaching Trigonometry." *AMM*. 54(1947): 36–37.

 V2   Vance, E. P. "The Teaching of Mathematics in Colleges and Universities." *AMM*. 55(1948):57–64.

 V3   Van Gross, J. A. "A New Ellipsograph." *SSM*. 22(1922):471–472.

 W1   Wagner, R. W. "A Substitution for Solving Trigonometric Equations." *AMM*. 54(1947):220–221.

 W2   Walsh, J. F. "A Paradox Arising from Integration by Parts." *AMM*. 34(1927):88.

 W3   Walsh, J. F. "A Rigorous Treatment of the First Maximum Problem in the Calculus." *AMM*. 54(1947):35–36.

 W4   Wang, C. L. "Proof of the Mean Value Theorem." *AMM*. 65(1958):362–364.

 W5   Ward, J. A. "A Function with Finite Discontinuities." *AMM*. 54(1947):162.

 W6   Ward, M. "A Generalized Integral Test for Convergence of Series." *AMM*. 56(1949):170–172.

 W7   Wavre, R. "Is There a Crisis in Mathematics?" *AMM*. 41(1934): 488–499.

 W8   Weaver, J. "Qualifications and Training of Teachers of Mathematics for Engineers." *Journal of Engineering Education*. 25(1935): 292–296.

W9   Weaver, W. "Note on the Teaching of the Principle of Mathematical Induction." *AMM*. 26(1919):190–191.

W10   Weil, A. "Mathematical Teaching in Universities." *AMM*. 61(1954):34–36.

W11   Weiss, M. J. "Algebra for the Undergraduate." *AMM*. 46(1939): 635–642.

W12   Wessman, H. E. "Mathematics in Civil Engineering." *Journal of Engineering Education*. 37(1947):378–382.

W13   Weyl, H. "Mathematics and Logic." *AMM*. 53(1946):2–13.

W14   Whitaker, H. C. "An Elementary Derivation of the Series for sin x and cos x." *AMM*. 7(1900):99–100.

W15   Whitford, D. E. and Klamkin, M. S. "On an Elementary Derivation of Cramer's Rule." *AMM*. 60(1953):186.

W16   Whitman, E. A. "The Film—a Triple Integral." *AMM*. 49(1942): 399–400.

W17   Whitman, E. A. "The Use of Models While Teaching Triple Integration." *AMM*. 48(1941):45–48.

W18   Willerding, M. T. "Is the Algebra Taught in College Really 'College Algebra'?" *MM*. 27(1954):201–203.

W19   Williams, K. P. "Note on Continuous Functions." *AMM*. 25(1918):246–249.

W20   Williams, K. P. "The General Equation of the Conic." *MM*. 16(1941–42):37–43.

W21   Wilson, L. *The Academic Man*. New York: Oxford University Press, 1942.

W22   Wollan, G. H. "What is a Function?" *MT*. 53(1960):96–101.

W23   Wood, F. E. "Derivation of the Tangent Half-Angle Formulas." *AMM*. 56(1949):103.

W24   Woods, R. *Analytic Geometry*. New York: The Macmillan Co., 1948.

Y1   Yates, R. C. "Classification of the Conics." *AMM*. 50(1943): 112–115.

Y2   Yates, R. C. "Folding the Conics," *AMM*. 50(1943):228–230.

Y3   Yates, R. C., and Nicholas, C. P. "Normal and Tangential Acceleration." *AMM*. 58(1951):255–256.

Y4   Young, A. E. "On the Teaching of Analytic Geometry." *AMM*. 16(1909):205–212.

Y5   Young, J. W. *Monographs on Topics of Modern Mathematics Relevant to the Elementary Field*. New York: Dover Publications, Inc., 1955.

Y6   Young, J. W. A. *The Teaching of Mathematics*. New York: Longmans, Green & Co., 1924.

Z1   Zant, J. H. "Critical Thinking as an Aim in Mathematics Courses for General Education." *MT*. 45(1952):249–256.

Z2   Zant, J. H. "Program for Determining the Mathematical Needs of Engineering Students." *MT*. 43(1950):91–94.